DESIGN AND ANALYSIS OF HIGH EFFICIENCY LINE DRIVERS FOR xDSL

DESIGN AND ANALYSIS OF HIGH EFFICIENCY LINE DRIVERS FOR xDSL

by

Tim Piessens

Katholieke Universiteit Leuven,
Heverlee, Belgium

and

Michiel Steyaert

Katholieke Universiteit Leuven,
Heverlee, Belgium

KLUWER ACADEMIC PUBLISHERS

BOSTON / DORDRECHT / LONDON

A C.I.P. Catalogue record for this book is available from the Library of Congress.

ISBN 978-1-4419-5424-4   e-ISBN 978-1-4020-2518-1

Published by Kluwer Academic Publishers,
P.O. Box 17, 3300 AA Dordrecht, The Netherlands.

Sold and distributed in North, Central and South America
by Kluwer Academic Publishers,
101 Philip Drive, Norwell, MA 02061, U.S.A.

In all other countries, sold and distributed
by Kluwer Academic Publishers,
P.O. Box 322, 3300 AH Dordrecht, The Netherlands.

Printed on acid-free paper

There is a theory which states that if ever anyone discovers exactly what the Universe is for and why it is here, it will instantly disappear and be replaced by something even more bizarre and inexplicable

. . .

There is another theory which states that this has already happened.

DOUGLAS ADAMS

Contents

List of Figures

List of Tables

Chapter 1

INTRODUCTION

THE recent evolutions in communication technology changed society in such a way that many people tend to state that we are experiencing a fourth industrial evolution. After the invention of the steam engine in 1764, the evolution to mass production in 1908 and the invention of automation around 1946, the advent of mass-communication changed economics drastically.

In 1970 , the telecommunications world consisted of voice-and character-oriented communications to mainframe computers. Voice was 'king', and there was little need for Digital Subscriber Line (DSL) technologies. Then came millions of personal computers, multimedia applications and eventually the Internet. In the early 1980s, the number of computers (including microprocessors in cars and appliances) exceeded the world population of humans, and in the mid-1990s the minutes of usage for digital applications in the public network exceeded voice. Broadband access meant business and the DSL technologies finally found a market that justified heavy investments into the research and development of a very complicated system for which in the end, cost would be a major issue for the economic success or failure of a certain solution.

The resulting Asymmetric Digital Subscriber Loop (ADSL) standards and the prospects of the Very high-speed Digital Subscriber Loop (VDSL) system have proved to meet the expectations for millions of broadband users. The implementations of these techniques had however a big drawback from a power consumption point-of-view, certainly at the Central Office (CO)-side, namely : its line drivers, the final building block between the modem and the telephone line, consume an enormous amount of power. The density of system boards is therefor no longer determined by component sizes but by thermal constraints.

The presented research activities are aimed at improving this building block, to design ADSL compliant line drivers with the highest efficiency ever reported. In this chapter, the importance of this work will be motivated and an overview of the content of this book is given, to guide the reader through this work.

1.   Motivation of the Work

Roman rhetoricians taught us that every enumeration should have exact three entries, and although the rather short title of this work : "High Efficiency Line drivers for xDSL" does not prospect an easy division into three parts, no exception to this rule will be made here. The motivation of the research activities presented in this work are thus threefold :

1 To prove with silicon that it is possible to design a highly efficient line driver for ADSL and to become state of the art in this field.

2 To advance the technology direction of integrating power modules in main-stream Complementary Metal-Oxide-Semiconductor (CMOS) technologies.

3 To build up knowledge in the field of non-linear system design and to develop models for the analysis of practical integrated non-linear systems and non-linear phenomena observed in the design of analogue building blocks.

1.1   xDSL Technologies

1.1.1   The Market

The advent of faster digital technologies, enabled a huge progress in the field of Digital Signal Processing (DSP) technologies. This enabled the use of complicated digital modulation schemes to get the most out of a telephone wire. These techniques were more and more getting used to connect telephone centrals, but the market opened with the advent of the Internet.

In recent years the Internet usage increased spectacularly, even in the post-.COM era. In 2002, the number of Belgian Internet users grew to 3.2 million from a mere 1.4 million in 2000 and 2.7 million Belgian surfers[1] in 2001 [De Graeve, 2002]. The same study revealed that more than 4 out of 10 of the Internet users are connected via a broadband connection like ADSL or cable. The number of broadband users has increased with 20% in the period from October 2001 to April 2003. Figure 1.1 shows the market evolution worldwide over the last 6 years. The exponential increase in demand for broadband access is linked with the higher bandwidth demand of regular sites by the use of more flashy designs and the increased downloading of multi-media content like mp3-music and even video. More than 30% of the newcomers on the web chose for broadband access since they have experienced a need-for-speed on the web.

---

[1]The cited study defines an Internet user as a person who uses the Internet more than once a month for surfing. People who only use email are not counted in these statistics.

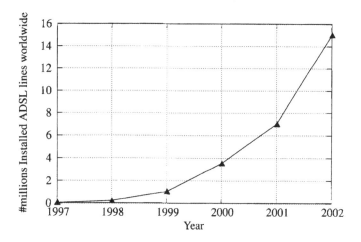

Figure 1.1: Number of installed ADSL-lines worldwide. source: Dslforum

1.1.2 The Investments

The Internet boom provided a huge market and the need to be first on the market triggered a huge investment by telecom companies to develop Digital Subscriber Loop (xDSL). This resulted in a tremendous bandwidth increase on the regular copper wire in a very short time. Consider the start of research on ADSL in 1991. At the end of 1995, Telecom Geneva already presented a first ADSL demo. The same can be said of VDSL. While research only started in 1995, a demo was presented near the end of 2000. So by an investment of 1 decade of DSL research 3 decades of speed was obtained, moving the copper wire form the kilo-bit to the mega-bit domain [Sevenhans et al., 2002].

1.1.3 Line Drivers : the gap in the xDSL system

These investments were justified by the large market potential and the return from the reuse of the old copper wire that connects over 700 million people worldwide. The speed by which all this was put together, did seemingly not allow sufficient feedback between telecom/system engineers and the analogue designers that needed to implement those systems.

This can be clearly seen in figure 1.2. To obtain this figure several ADSL chip-sets from different companies were regarded and the relative sizes of the power consumption of the analogue and digital building blocks and the line driver were calculated. By averaging out over several companies, a representative figure for the state of present products can be given. The line driver took almost 60% of the total power budget. The exercise was redone for the next generation solutions of the same vendors. The relative size of the total pie represents the relative decrease in total power consumption. From this pie-chart

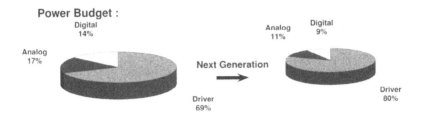

Figure 1.2: Evolution of the total power budget for an ADSL-chip set

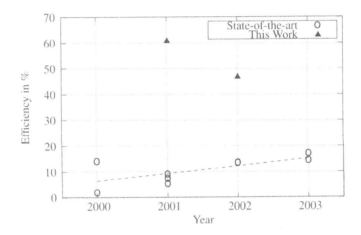

Figure 1.3: Reached efficiency in recent state-of-the-art line drivers

one can learn that as expected digital and analogue power consumption drops going to a next generation. This is due to the fact that the analogue and digital components can nowadays be fully integrated in mainstream CMOS technologies [Conroy et al., 1999, Sands et al., 1999, Cornil et al., 1999b]. Due to the decrease in channel lengths and the increased matching, digital and analogue building blocks will consume less power. The line driver, however, does not seem to benefit from this evolution. This is due to the specific modulation that is used to reach high bandwidths in ADSL. ADSL signals have a noise-like look with several voltage peaks. These large Crest Factors (CFs), meaning the ratio between the maximum voltage and the rms signal voltage, render traditional class AB power amplifiers to be low efficient line drivers.

The gap between the maximal allowed power consumption and the present state-of-the-art can be clearly seen in figure 1.3. The xDSL line drivers from competitors presented at the International Solid-State Circuits Conference

Table 1.1: Overview of the most important driver architectures in relation to the maximum number of lines per 500 cm^2 board.

Lines per 500 cm^2 board	24	48	72	96	120
Class AB 740 mW		NOT			
Class G 400 mW				NEBS	
ΔΣ Class D 200 mW		NEBS			COMPLIANT
SOPA 100 mW		COMPLIANT			

(ISSCC) are depicted and a trend-line is drawn through them by a dashed line. As a teaser, the presented work is put on the same figure.

1.1.4 Breaking The Last Barrier in Wire-line Communications

The research presented in this book is dedicated to create high efficiency line drivers for xDSL. The importance of high efficiency can be easily observed by taking a look at the Network Exploitation Board Specifications (NEBS) norms. The NEBS prescribe the maximal amount of power that can be dissipated on a 500 cm^2 board. If the efficiency of the line drivers is taken into account, the power dissipation of the line driver will fix the maximum number of lines that can be served by a single board. The results of these calculations are depicted in table 1.1 [Sevenhans et al., 2002]. Board density nowadays is no longer limited by component size but by thermal limitations. Table 1.1 also shows that traditional structures do not suffice. So a novel topology : the Self Oscillating Power Amplifier (SOPA) is presented here.

By the research of a switching type line driver with a very good linearity, the power bottleneck of figure 1.2 can be broken. The goal is to design highly efficient line drivers that do comply with the ADSL specifications.

1.2 Power Amplifiers in CMOS

Another important motivation for this work is the increased interest in the design of power amplifiers and other power building blocks in mainstream technologies. While analogue designers are getting very near the full integration of analogue front-ends in pure digital CMOS, the power building blocks have been separate chips. For mass-production as cellular phones, the integration of the power amplifier could lead to a lower production cost and even smaller handsets [Mertens and Steyaert, 2002].

A major issue when integrating power amplifiers in CMOS technology is the fact that when going to deeper sub-micron technologies, the supply voltage drops for reliability reasons. Since the output power is coupled with the maximum voltage swing by $P = V^2/R$, This makes it very hard to design line drivers in a CMOS-technology.

With this work, a contribution has been made to this field, by choosing to implement the SOPA in a mainstream CMOS technology. Throughout the book, much attention will be paid on the restrictions of submicron CMOS and how to handle them. Also scaling laws for a SOPA in digital CMOS technology will be derived, so one can more or less predict the possibilities of this technique in future CMOS technologies. The concepts and techniques presented in this work are also applicable for other power building blocks like high efficiency DC/DC converters or Radio-frequency (RF) power amplifiers [Su and McFarland, 1998].

1.3   Non-linear System Design

Engineers like to model the world as a linear space, for many mathematical techniques exist to analyse a linear model. Moreover a linear system obeys the superposition principle, so the favourite technique of 'divide and conquer' can be applied. By splitting up the system in smaller parts, the complexity can be handled. For non-linear systems however, the superposition principle does no longer hold and the system needs to be analysed in its completeness.

Another drawback of non-linear systems is that they tend to use an enormous amount of CPU-time for numerical simulations. This make numerical simulations hardly usable in the system design phase of many non-linear systems, certainly if design centring is required. An analysis technique, based on straight forward calculations or calculations with limited iteration steps is therefor necessary to characterise the system. The feedback from these models provides insight to the designer and can steer the design to his specific needs.

The SOPA amplifier is a continuous time switching line driver. Therefor a hard non-linearity will be present in the system. The third and last goal of this work is to develop an analysis system to fully predict the output of a hard non-linear SOPA-like system. Although these techniques should eventually lead to design plan for the SOPA line driver itself, the presented techniques are made as broad as possible so they should be usable for other non-linear systems like Phase-locked Loops (PLLs),DC/DC-converters, etc..

2. Organisation of the Book

The book largely follows these three major motivations and tries to span the complete design process from a problem study over system analysis, concept development to real system implementation and proof by measurements on silicon.

Chapter 2 will investigate the required specifications of an xDSL line driver. Since these requirements are a direct consequence of the used channel, the history of the telephone twisted pair is shortly described. The xDSL standard will be placed in its historical context. From this short overview, one can more easily understand why people want to use a channel that was originally designed for 3 kHz limited voice communication for broadband data transport. The properties of the channel are the next subject in this chapter. This knowledge will aid the understanding of the specifications and clarifies the hows and whys of Discrete Multi-Tone modulation (DMT). The statistics of DMT-signals will determine eventually the capabilities of present line driver solutions. A brief overview of the most important reported structures are given and their maximum performance will be critically discussed. This will lead to the conclusion that a novel self-oscillating switching structure is necessary to provide a true highly efficient line driver.

Since the proposed SOPA structure is a hard non-linear system, traditional (linear) control theory cannot be used. In chapter 3, the mathematical tools to analyse a SOPA structure will be discussed. The used describing function method provides a quasi-linearisation of the system. Since this approximation of the real system's behaviour introduces errors, it is of the utmost importance to understand the backgrounds of this technique to draw realistic conclusions from the analysis.

After describing the tool-set, the real work can get started. In chapter 4, a general model for the SOPA-concept is presented and the analysis techniques are applied. The major issue is the limit cycle oscillation. Techniques to calculate its frequency and amplitude will be presented. The limit cycle self-oscillation will act as natural dither in the non-linear system. This dithering linearises the system and provides some self-adaptivity properties which enable a robust, linear line driver. The possible distortion and Missing Tone Power Ratio (MTPR) levels will be calculated and straight-forward equations will be derived to map the system's structure on the output characteristics like third order distortion and signal bandwidth. Another important feature of a SOPA line driver is that it uses oscillator pulling into synchronisation to reduce output filtering requirements and power consumption. The synchronisation modes and which one dominates for several line conditions will also be derived in this chapter. The last part of this chapter introduces the improvement of adding noise-shaping techniques to the SOPA system. These higher order SOPA amplifiers will enable linearity levels as required by the ADSL-specifications.

While the analyses of chapter 4 are all on the system level and are pure mathematical modelling techniques, a bridge to real-life implementations needs to be built. This is covered in chapter 5. This chapter consists of two major parts : the design plan synthesis and some Computer Aided Design (CAD) techniques that were developed to support this design plan. The design plan starts from real life limitations like process technology and transformer requirements to adapt the system specifications into design parameters. For the process technology CMOS scaling rules are derived so that the feasibility can be investigated before hand. Also do these scaling laws provide a vision on the future possibilities for integrating a SOPA in CMOS. After the feasibility is confirmed, the design plan takes the results from the analysis methods from chapter 4 and orders them in such a way that the system's specification are converted into building block specifications without too many iteration steps.

In the next part of chapter 5, some CAD-techniques are presented to support non-linear design and the presented design plan. Since this research is design focused, the discussion is focused on numerical stability and accuracy and design time speed-up by integrating several components into a single environment. This means incorporation of the behavioural modelling effort of chapter 4 in a set of design tools. The behavioural modelling implementation provides an improvement of 5 orders of magnitude in simulation speed-up compared to behavioural modelling in ELDO and 3 orders of magnitude on a .m-file Matlab implementation.

Modelling efforts and system concept simulations are as good as the quality of the models. A real implementation of the presented system needs to prove the concepts. In chapter 6 two different implementations in a mainstream .35 μm technology are presented. The most important design issues are given and the obtained measurements are described and compared with the results from the system analysis. The first design is a zeroth order SOPA amplifier. Measurements prove its full compliance to less performing ADSL-Lite (G-Lite) specifications and this for a measured efficiency of 61%. To cope with the limited linearity of the zeroth order design, the next design needed to be of higher order. A third order design has been chosen and the goal was set more ambitiously to design a multi-standard xDSL line driver. This implies a line driver with the stringent linearity specifications of ADSL and the high bandwidth of VDSL. The processed SOPA had a bandwidth of 8.6 MHz, for a MTPR of 56 dB. In this way it complied to the ADSL and VDSL power spectral density mask. This was reached with a total efficiency of 47% for an ADSL signal with a CF of 15 dB.

In a last chapter the major conclusions from this research work are presented and some possible future improvements will be discussed in short.

Chapter 2

XDSL LINE DRIVERS:
SIGNALS, SPECIFICATIONS
AND TRADITIONAL SOLUTIONS

THE telephone has been since its introduction the most important telecommunication medium. More than 700 million households are connected to this immense network. More than a century long the telecom companies have dug up the streets and have put in tons of copper to connect these subscribers. It may thus not surprise us that efforts have been put into the reuse of all this cable to deliver broadband access to the Internet to those subscribers. In a first section the history of data communication over the telephone network is sketched. Since compatibility was mandatory between all systems, a knowledge of telephone history is mandatory to fully understand the various inherited constraints to design an analogue front end for xDSL.

The twisted pair cable used for telephony is a very poor channel for high bit-rate communication. In the second section, its basic properties will be explained. The channel losses and unpredictable behaviour at high frequencies will require special modulation techniques to allow the high bit rates of xDSL. Cable impairments like bridged taps, cross talk and radio frequent interference will limit the throughput through the channel.

The maximum channel capacity has been predicted by Shannon. this theoretical limit holds in its definition a clever possibility to fully exploit the channels capacity. By discretising the bandwidth and load every carrier with a bit rate which is proportional with the measured signal-to-noise ratio in that frequency bin, the full capacity is usable. By the rise in processing power and DSP techniques, the implementation of DMT became possible. A third section will discuss the basic properties of DMT-modulation and its consequences for the design of an Analogue Front-End (AFE) in general and more specific the line driver. From this the major xDSL specifications can be derived.

The high CF of DMT signals proves to be the major difficulty to create a high efficiency line driver. Therefor the line driver is the most important power consumer on a CO line board and puts the limit on the amount of lines that can be installed on the same board. Density is no longer limited by spatial considerations but by thermal limits. In a fourth section the challenge for designing an xDSL line driver are further elaborated.

in the last section, traditional solutions for constructing a line driver are elaborated. The mostly used class AB line driver provides a very linear power amplifier but its power consumption is extremely high. A logical improvement is the class G and eventually class H topology. By toying with the supply of a class AB line driver, higher efficiencies can be reached. A switching type line driver, however is the way to go for decreasing the power consumption to a very minimum. The basic design issues will be touched and a novel self-oscillating type line driver will be introduced. This SOPA technique will be the subject of the rest of this book.

1.   Broadband Communication Technology

1.1   The Beginning

Although it is assumed to be a recent technology, digital data communication is a very old technology. The birth of digital communication can be dated May 24th, 1844 with the invention of the telegraph by Samuel Morse. Data was encoded in the form of dots and dashes like the 0 and 1's of nowadays communication systems. The Morse code has an inherent data compression, since it was aimed to be operated by hand. This limited the 'bit-rate' to 4 to 5 dots and dashes per second. On top, this bit-rate could only be achieved by professionally trained telegraph operators. From a communication point-of-view, this is the major bottleneck for real data transmission.

A first approach to widen the applicability of the telegraph for data transmission was the invention of the printing telegraph in 1874. This invention automated the transcription of the data in an automated fashion. The goal was two-fold : firstly, by developing a typewriter style keyboard interface to the telegraph, non-professional people were able to operate the new telegraph. Secondly, the automation of transcription enabled a faster transmission than that could be achieved by human operators. The French inventor Jean-Maurice-Emile Baudot developed a five bit code to represent the characters to be transferred. This code dropped the inherent data compression of the Morse code, but enabled further automation and time division multiplexing, another invention by Baudot. To this day, the speed of serial communication, Baud rate, is still named after this invention. After several improvements, made by an Englishman named Donald Murray, the tele-typewriter made its way to the public communication network and gradually replaced the traditional telegraph.

The second approach was an early attempt to use some sort of frequency multiplexing on the telegraph lines to increase the throughput. A first attempt was made by Helmholtz. Helmholtz, who was in research of the physical basis of music, had built a setup which used electro-magnets to set a tuning fork. Several inventors picked up his idea. If an electrical wire was able to make a tuning fork sing, would it be possible to make a musical telegraph. Since a melody has a very large information content, it would in this way be possible to transmit many messages at the same time. The German inventor Phillip Reis, was influenced by the early work of the French inventor Bourseuil who described a method to convert speech in electrical vibrations. In 1854, he wrote : " Speak against one diaphragm and let each vibration 'make or break' the electric contact. The electric pulsations thereby produced will set the other diaphragm working, and [it then reproduces] the transmitted sound. " [Lienhard, 2000]. Reis' telephone used a diaphragm who didn't 'make or break'[1] the contact but drove a rod with varying depths into an electrical coil. The continuously varying current, made a better representation of sound, which allowed a clearer transmission of speech. However, the diaphragms of Reis were unstable and its experiments were inconsistent over various repetitions of the experiments. It was up the advent of an inventor who combined the knowledge of acoustics with the knowledge upon the subject of electricity that a workable telephone could be created [Casson, 1910]. It was the Scottish American Alexander Graham Bell, who devised the first working telephone in 1875. Although his invention was ridiculed at first or seen as merely 'a scientific toy', its importance in the history of telecommunication cannot be exaggerated. The main advantages which made the telephone a worldwide success were mentioned in a circular made by Bell and his partners in 1877 :

- No skilled operator is required, but direct communication may be had by speech without the intervention of a third person[2].

- The communication is much more rapid, the average number of words transmitted in a minute by the Morse sounder being from fifteen to twenty, by telephone from one to two hundred.

In the course of history, many people improved the telephone. The invention of the microphone by David Hughes, with further improvements by Thomas Edison who introduced the carbon granule transmitter, made the telephone into an instrument that was much more sensitive than Bell's Aluminium alloy diaphragm. The transmitter further grew into a complete artificial ear and mouth containing a large number of components and a salt-spoon of carbon. These

---

[1] A careful reader will recognise an ironic twist of history. Although Bourseuils idea to represent sound by a pulsed current gives a complete distorted representation, the further development of line drivers will automatically lead to a 'make or break' class D implementation.

[2] Of course this claim was made before the big explosion in telephone use and the need for (at first) human operated switching boards.

components however needed to be powered, in contrary to the original telephone. Originally they were powered by small batteries installed in every telephone set. This changed in 1896 by a revolutionary change in switchboard techniques. This was mostly due to the invention of the invention of the 'girlless, cuss-less[3]' telephone system by Almon Brown Strowger. His invention provided the dialling services that are still known today. To power the dialling service at the customer premises, the line is powered from the CO. This also meant better signalling and better talking, since the supply was now well controlled. It reduced the cost of batteries and put them in the hand of experts. Furthermore, it established stability and uniformity, enabling telephone service providers to upgrade their systems almost independent of the equipment of its users. Line powering is still one of the major issues in present modem design as will be explained further in this chapter.

1.2   Cable Technology

Not only the apparatus as such, but the complete telephone system went through a revolutionary change. Equally important to the telephone set is the cable used to conduct the information. In the early days of telephony, people leased telegraph lines to connect their telephones. The steel cables, however, made noisy communication. The first users of the telephone were surprised to hear 'ghostly sounds' through their telephone sets. Since the telegraph lines used a single wire and the loop was closed by ground and the earth acts as a gigantic magnet, the grounded loop picked up various magnetic signals. In 1883, J. J. Carty used two wires to connect a new line between Boston and Providence. The first 'quiet line' was conceived. This meant doubling the amount of wires and thus the investments for the telephone companies.

By the use of two wires the lines became quieter but only for a short time. As business bloomed and more people subscribed to the telephony system, the lines again produced unexpected noises. This was due to crosstalk between parallel telephone lines, due to inductive coupling of the wires. This problem became more eminent in big cities. Due to the advent of the skyscraper, wires needed to be put into the ground. Many wires had to be put into large trunk closely together. Crosstalk seemed to be inevitable. In 1881, Graham Bell came up with an easy but drastic solution to this problem : he invented the twisted pair. By twisting the wires, the disturbing signal becomes common mode to the wires and is in that way heavily suppressed. Although the twisting schemes[4] evolved, present telephone wires are still built on the same principle.

---

[3]The biggest improvement in the telephone exchanges before 1896 was the replacement of boy operators by girls. This not only reduced noise but improved switching speeds from 5 minutes to 20 seconds.

[4]Phone wires generally have a twist every 2 to 6 inch which vary in their course to reduce cross talk.

Another cabling issue is the choice of material. The original steel wires were strong but had a very high resistivity degrading the signal quality. Copper seemed to be the best choice from an electrical point-of-view but was too soft and too weak. It couldn't even carry its own weight. This problem was solved by Thomas B. Doolittle by hard-drawing copper wires. All problems seemed to be solved to use copper wires, except its price. The cost of the copper wires was so enormous that fully one fourth of all capital invested in the telephone business has gone to the copper mines. To lower the resistance of long copper wires, they had to be made fairly thick, which was ruinous for the telephone companies. The problem lies in the propagation constant of a transmission line. If a telephone is represented by its lumped transmission line model, the propagation constant $\overline{\gamma}$ can be calculated as follows :

$$\overline{\gamma} = \sqrt{(R + j\omega L)(G + j\omega C)} \tag{2.1}$$

If the following condition

$$\frac{L}{R} = \frac{C}{G} \tag{2.2}$$

is fulfilled, the complex propagation constant becomes :

$$\overline{\gamma} = \sqrt{RG}\left(1 + j\omega\frac{L}{R}\right) \tag{2.3}$$

This propagation constant has two important properties : its attenuation is independent on the signal frequency, while its phase component is linearly proportional with frequency, which is equivalent with a loss-less line with same inductance L and capacitance C. So, the dispersion of the line is zero and the wave-forms are only attenuated and not distorted. If the resistance of the wire is too high to fulfil condition (2.2), speech will become deformed and ultimately communication would become impossible. This problem was solved by Pupin in the 1890s. By introducing discrete load coils in series with the line at fixed distances, the line inductance is virtually increased. These coils, however, are extra pass-bands filters in the transmission lines. The bandwidth of a long telephone line is limited to 3.6 kHz by the introduction of these load coils. Figure 2.1 illustrates this effect. In this figure the transfer function of a 16 km long unloaded and loaded 24AWG wire is depicted. Both wires are terminated with a 100 Ω load resistance. The loaded wire is loaded every 1.8 km with a 88 mH coil.

1.3 Voice-Band Modem Technologies

In the 1950s, people started experimenting sending digital data over the copper wires. Due to the Pupin-coils, early modems suffered the limited bandwidth of the Public Switched Telephone Network (PSTN), which was only

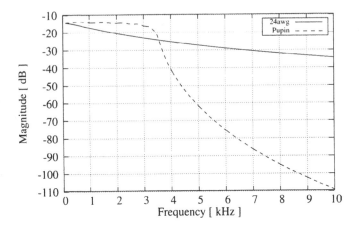

Figure 2.1: Transfer functions of a loaded an unloaded 24AWG line to demonstrate the effect of Pupin coils.

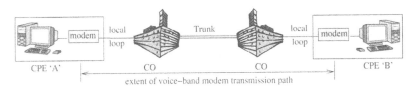

Figure 2.2: Voice-band modem reference model

focused to transmit speech. The digital signals that had to be put on the line needed to be modulated since the PSTN does not convey frequencies below 200 Hz. The word modem is constructed to emphasise the modulator-demodulator function of these voice-band modems. One of the first voice-band modems, AT&T's Bell 103, was used for full-duplex, asynchronous type transmission at 300 /s, which was very little, but limited by the PSTN.

Figure 2.2 shows the reference model of the voice band modem. The digital data was converted into an analogue signal with limited bandwidth. It is then transferred over the analogue telephone network. The PSTN can be considered to consist of three major parts. A first part is the local loop between the Customer Premises Equipment (CPE), the installation at the callers side, and the CO, the exchange the customer is connected with. By a cascade of switchboards, this CO is then connected to the CO closest to the called customer. When a connection is established a direct connection between caller A and B is established. This wired connection is only used by those two callers. Since the trunk connection between the COs could be fairly long, it is most likely loaded. Although the network had analogue amplifiers, mostly at the

Figure 2.3: Reference model of the modem with PCM trunks

COs to boost the signal, the signal deteriorated gradually. The cross-talk in the large trunks limited the data throughput. Leased lines, without Pupin-coils could offer higher bit-rates, but due to incompatibility with international phone networks, these modems hardly hit the market.

In the 1970s, with the increase of telephone traffic, the trunks connecting were congesting. A first digitisation of the old trunk cable was mandatory to increase the throughput of data between the central offices. Note that the telephone companies had invested enormously in copper wire. So they were not eager to dig new connections between their COs. The trunks were transferred into Pulse Code Modulation (PCM)-trunks. Voice calls were digitised to PCM at a rate of 64 k/s. By the use of digital repeaters, the copper wires could carry 1.5 to 2 M/s. Twenty-four to thirty-two voice channels were one-way multiplexed on one twisted pair. This was a first example of the reuse of the old copper wire for high bandwidth applications by using digital techniques. The transmission quality increased since PCM transmission is free of analogue noise. This however happened transparent to the user. But an important step was taken : the long line connection was cut into two local loops and a transparent network between the COs. No direct connection was needed and the copper between the COs was utilised by more than one user. The transmission model is given in figure 2.3

Another thing happened: the arrival of digital signal processing. In 1981 the V.32 modem introduced Trellis coding and took a bold step by using in order to transmit information in the same frequency band. The signals however remained analogue signals within the frequency band of the speech signals. The limitation for data transmission was transferred from the channel to the Coder-Decoder (CODEC) function at the CO. The fundamental limit for data transmission was 64 k/s, if only one channel per customer could be used.

In the 1980s, telephone companies increased there investments in their backbone network. The coax and copper trunks were replaced by optical fibres with the sonet system. The connection between the COs was now able to provide high-speed packet switched data-transfer for applications like the Internet. The bottleneck was shifted again to the local loops. The network topology is shown in figure 2.4. At first the modems were still voice-band modems, that due to the shorter loop lengths and less interference in the local loop, could utilise

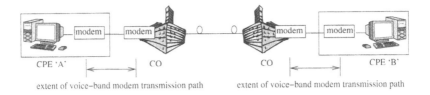

Figure 2.4: Reference model of the modem with the fibre backbone.

the full voice-band bandwidth. In the 1990s the local loops were replaced by PCM-trunks. The V.90 56 k/s PCM-modems appeared. But their achievable transmission rate is limited on transmitted power and line impairments. From this point it is also possible to remain in the digital domain from end to end if the CO is able to map the transmitted PCM symbol into a digital signal that can be send over its backbone network. Quantisation noise is suppressed in this way.

An overview of the different voice-band modem technologies is given in table 2.1. Note that although the V.90 is the last modem included in this list, it is the V.34 that is the last to assume the line is analogue. The V.90 can be regarded as a transition technology.

Table 2.1: Overview of different voice-band modem standards

standard	date (ratified)	speed (bps)	half (HDX)/ full duplex (FDX)	PSTN/ private	modulation
V.21	1964	200	FDX(FDM)	PSTN	FSK
V.22	1980	1200	FDX(FDM)	PSTN	PSK
V.22 bis	1984	2400	FDX(FDM)	PSTN	QAM
V.23	1964	1200	HDX	PSTN	FSK
V.26	1968	2400	HDX	Private	PSK
V.26 bis	1972	2400	HDX	PSTN	PSK
V.26 ter	1984	2400	FDX(EC)	PSTN	PSK
V.27	1972	4800	HDX	Private	PSK
V.27 bis	1976	4800	HDX	Private	PSK
V.27 ter	1976	4800	HDX	PSTN	PSK
V.29	1976	9600	HDX	Private	QAM
V.32	1984	9600	FDX(EC)	PSTN	QAM
V.32 bis	1991	14400	EC	PSTN	TCM
V.34 (V.fast)	1994	28800^a	EC	PSTN	TCM
V.90	1996	56000^a	EC	PSTN	PCM

[a]These figure refers to the maximum available data-rate

1.4 Digital Subscriber Loop Modems

1.4.1 The early beginning : ISDN

The start of DSL[5] is given in the early 1980s with the introduction of Integrated Service Digital Network (ISDN). The specifications were first conceived in 1976. The ISDN vision was very ambitious : to construct a global network for data communications and telephony. The effort to develop ISDN spanned a decade. The total development cost is estimated to be over $50 billion [Starr et al., 1999]. ISDN was focused on telephony services and lower-speed packet switched data. This focus ultimately became a major weakness. ISDN networks were poorly suited for the high-speed packet switching and long holding-time sessions that characterise Internet access. Since ISDN provides a digital channel based switched connections, the interfaces at the CO and CPE differ drastically. To refer to this change, the term modem is exchanged by the Line Termination (LT)/Network Termination (NT) terms.

Basic Rate ISDN (BRI) transports a total of 160 k/s of symmetric digital information over loops up to approximately 18 kft[6]. This reach was later on extended by the introduction of mid-span repeaters or the introduction of more advanced digital signal processing. The Extended-Range BRI uses techniques like Trellis coding to permit 160 k/s over lines up to 28 kft. For backward compatibility, the Extended-Range BRI's LT and NT needs to be followed by a converter to allow the 28 kft range. Figure 2.5 shows an overview of the different basic ISDN network architectures.

BRI uses Pulse Amplitude Modulation (PAM) to modulate its data. A four-level pulse (a quat) represents two binary bits, hence the name 2 Binary, 1 Quaternary (2B1Q). BRI is the first consumer or residential DSL. The 160 k/s is channelised as two 64 k/s voice or data 'B' channels, one 16 k/s control 'D' channel, and one 16 k/s channel for framing and line control. The 'B' channel may be circuit switched or packet switched. The 'D' channel carries signalling and user data packets. An embedded operations channel and indicator bits are contained within the 8 k/s overhead. The embedded operations channel conveys messages used to diagnose the line and the transceivers. The indicator bits identify block errors so that transmission performance of the line may be measured.

---

[5]Since ISDN is considered as the earliest implementation, the term DSL is used to describe these techniques. If the more advanced DSL members are meant like ADSL, VDSL, etc., the term xDSL is used.

[6]Since the origin of telephony lies in the US, imperial units are still used today to characterise loop lengths. 18 kft corresponds with approximately 5.5 km

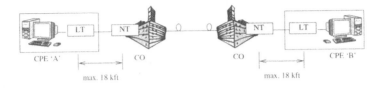

(a) Basic Rate ISDN

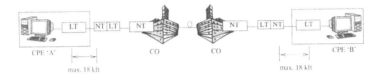

(b) Basic Rate ISDN with mid-span repeater

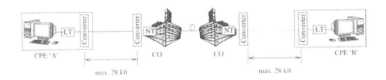

(c) Extended-Range Basic Rate ISDN

Figure 2.5: Overview of major ISDN topologies

1.4.2   The further evolution to xDSL

Note that the evolution in access techniques was only possible due to the evolutions in the inter-office trunks. In 1986, the early concept definition of High-speed Digital Subscriber Line (HDSL) started. When the telecom operators ceased to use the twisted pair trunks (T1/E1 transmission) between their COs, these lines were freed to be used as private lines between CO and CPE. T1/E1 transmission operate over the existing telephone wires, but at a large cost for special engineering, loop conditioning (removal of bridged taps, see section 2.2.3.1, and loading coils), and splicing of apparatus cases to hold the repeaters that were required every 3 to 5 kft. The transmission methods used foe T1/E1 lines placed high levels of transmit power at frequencies from 100 kHz to above 2 MHz ; this required the segregation of T1/E1 lines into separate binders to diminish cross-talk.

HDSL was aimed to provide a plug-and-play transmission system that could quickly and easily provide 1.5 to 2 M/s over most subscriber lines. Plug-and-play meant HDSL should be applicable on most telephone lines, thus should be able to cope with bridged taps[7]. This is accomplished by the inclusion of more extensive diagnostics features, including Signal-to-Noise Ratio (SNR) measurement. Also HDSL causes less cross-talk since it needs to transmit less power than T1/E1 transmission. HDSL's benefits are mostly due to the elimination of mid-span repeaters. The total cost of a repeater is mostly formed by the cost during operation. A repeater failure means a field service visit. They are also mostly line powered; this requires a special line feed power supply at the CO. Most of the power fed by the CO power supply is wasted due to loop resistance.

HDSL provides two-way 1.544 or 2.048 M/s transport over telephone lines up to 12 kft. The HDSL systems use two pair of wires, with each pair conveying 768 k/s of payload in both directions. Thus the term *d*ual duplex is used to describe HDSL transmission. The standard also opens three wires and single wire (SHDSL) operation. The modulation scheme is echo-cancelled hybrid 2B1Q. The bit rates reported were maximum bit rates for very short loops. In a second generation (HDSL.2) more improved modulation schemes that used 16 level PAM. The spectra of HDSL mostly remained symmetric which is favourable for normal data traffic, but is unusable for more recreative applications.

A possible entertainment application marketing experts in the early nineties envisioned was video-on-demand. The development of new compression standards for video like MPEG, Video-on-demand needs a high downstream bit rate (from CO to CPE) but requires a very limited upstream (from CPE to CO) data-rate. Video-on-demand never fulfilled its prospects. This was mostly due to :

- Large investments in video equipment were necessary

- A major upgrade of the backbone network was necessary. To offer a reasonable quality-of-service, multiple video servers needed to be included in the network, as close as possible to the customers.

- MPEG-1 digital video offered a video quality that was too poor.

- The setting of a business needed the cooperation of a major telecommunication firm, who owns the infrastructure, with a broadcast company, who owns the rights on the media. This is a major economic bottleneck.

The death of its major application didn't cause ADSL to die, since something else happened : the Internet. With the big boom of Internet related companies

---

[7]HDSL cannot cope with loaded lines. Since for service reasons, lines were getting shorter by the installment of local exchanges, loaded lines were getting a rarity in the local loops.

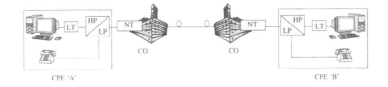

Figure 2.6: Topology of an ADSL network

and applications, the demand for high-speed access to the Internet dramatically increased. Since telephone companies had invested enormously into wiring for more than a century, a technology that could make broadband access available, through the twisted pair would turn the tons of buried copper into a gold mine.

The deployment topology of a generic ADSL network is depicted in figure 2.6. The ADSL technology was aimed at the non-professional user, so a new subscriber should be able to use ADSL out-of-the-box and without modifications on its existing communication network, i.e. the Plain Old Telephone Service (POTS) telephone system. The local loop conveys simultaneously the following signals through one pair of wires :

- Downstream bit rates of up to 9 M/s

- Upstream bit rates of up to 1 M/s

- POTS, i.e. analogue voice signals.

By the use of frequency division multiplexing without overlap, the splitter function that separates POTS with the ADSL -signals consist of mere low- or high-pass filters. Line attenuation and crosstalk make up the canonical impairments for defining DSL performance. Table 2.2 shows the downstream rates that can be achieved on 24 American Wire Gauge (AWG) wire, when a reasonable crosstalk is assumed. Note that about 80% of the lines in the United States are shorter than 18 kft [Maxwell, 1996]. The region called the *carrier serving area* extends to 12 kft, and encompasses about 50% of the lines in the United States.

The last three rates of table 2.2 fall under the VDSL system. This member of the DSL family is an extension of the ADSL technology to higher bit-rates. At such high rates, the loops must be so short that optical fibre will be used for all but the last few thousand feet (i.e. the last mile). VDSL will primarily be used for loops fed from an Optical Network Unit (ONU), which is typically located less than a kilometre from the customer. Few VDSL loops will be served directly from a CO. Optical fibre connects the ONU to the CO. VDSL is intended to support all applications simultaneously : voice, data and video. Ultimately VDSL would support High Definition Television (HDTV) and high-

Table 2.2: Achievable bit-rates for an xDSL connection versus loop length

	downstream bit-rate	loop-length
ADSL	1.5 M/s	18 kft
	2.0 M/s	16 kft
	6.0 M/s	12 kft
	9.0 M/s	9 kft
VDSL	13.0 M/s	4.5 kft
	26.0 M/s	3 kft
	52.0 M/s	1 kft

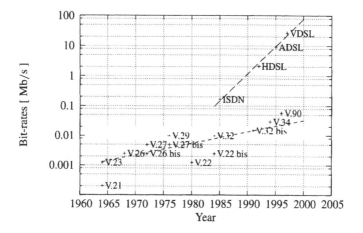

Figure 2.7: Maximum downstream bit-rate versus year of introduction for various twisted-pair data communication technologies

performance computing applications. Symmetric application of VDSL will provide two-way data rates up to 26 M/s that will be attractive for business sites where fibre-to-the-building is not justified.

The way these bit-rates are achieved will be more elaborated in the following sections; but while reading the techniques involved with xDSL, one should always keep in mind the legacy of POTS. One of the big advantages of xDSL is its compatibility with POTS, nevertheless it costs flexibility and increased building block specifications. The knowledge of the inherited system, helps understanding the choices made designing an xDSL system.

1.4.3   Overview of twisted-pair techniques

Figure 2.7 shows the achieved bit-rates versus the year of introduction for different twisted-pair data communication technologies. The improvement from voice-band technologies to DSL-technologies concerning bit-rates is tremendous.

More striking however, should be the speed by which the DSL-technologies improves. In one decade of xDSL design the copper-wire transmission went from the kilo-bit to the mega-bit range. The trend-lines through the different technologies are explicitly drawn in figure 2.7 by short dashes for voice-band technologies and long dashes for DSL-technologies. The growth-rate in log-arithmic scale of the DSL evolution is more than 4.6 times the growth-rate of the voice-band modems.

1.5   The Competition

The competition to DSL to carry high-bit-rate data transport to the home can be summarised by the three alternative carriers : fibre, coax and wireless technologies. While in the late 1980s, it was popular belief that fibre optics would dominate data transport in 'a few years'. Fifteen years later, fibre still is marginal to connect the last mile. Why fibre didn't meet the expectations is just a matter of economics ; there does not exist a major application to create the demand and thus the market for broadband access to justify the large investments in cabling and logistics to make a totally photonic world. Since next to the future possibility to have video-on-demand, the Internet is the only driver for broadband access. Due to the limited backbone capabilities, there is no need for higher speed in the local access points.

A recent market study [In-Stat/MDR, 2002] reported a growth of the to-tal number of broadband subscribers to pass the 30 million mark. World-wide, the portion of DSL subscribers finally took the main portion with 17 million subscribers. In the United States, however, coax cable communica-tion still takes the lead with 7.12 million subscribers next to 4.6 million DSL-subscribers. This is mostly due to the "Triple Play" bundled service packets of voice, video and high-speed Internet access - a marketing package that DSL service providers can rarely match. Another economic element in favour of cable access is, when video-on-demand will finally hit the street, cable compa-nies are better equipped to offer these kind of services. Since cable companies are closer connected to broadcast companies, they will have better access to the copyright protected movies and media. A non-technical argument in favour of DSL is the so-called life-line support. Since a telephone is line powered, it can still be used for emergency calls if the power at the customers side is down. This life-line support is in some countries enforced by law, limiting the use of coax for telephony services.

From the technical side, both technologies have matured and both have their pro's and con's [Frenzel, 2001]. Some technical considerations are :

- xDSL is limited by cross-talk and radio frequent interference. Since coax cable is shielded, these problems do not occur. The coax cable intrinsically is better suited to provide high bit-rate data access.

- Like DSL, coax cable also carries the burden of history. Most cable systems are a Hybrid Fibre/Coax (HFC) arrangement with a main fibre trunk cable distributing the content to many coax cable nodes. These feed the drops to each home. Since the system was originally designed for broadcast, the final coax cable runs via many households. The channel needs to be shared amongst neighbours. This will limit the bandwidth during rush hour.

- Coax cable bandwidth is organised in adjacent 6 MHz channels designed to carry TV-signals. Coax bandwidth is as much as 860 MHz, so over 120 channels are available. However only a few are allocated to data applications.

- Cable systems are very asymmetrical. They can download signals at up to 27 M/s. Upstream channels are only 2 MHz wide and located in noisier regions of the cable's spectrum, namely the 5 to 40 MHz segment.

- Maximum theoretical upload speed is 10 M/s but most cable companies limit up-link speed to between 256 and 384 k/s. Since this up-link speed needs to be shared by all the users on the same cable, it can be a serious limitation to the growth of the network.

- Since cable companies, in contrast with telephone companies, originally only did one-way communication (broadcasting), most cable facilities could only handle one-way communication. In 1997 only 10% of the cable customers were connected via a two-way cable facility. Cable companies are upgrading their facilities at a high pace, but only at selected areas, since the costs need to be justified by a reasonable future income.

- Cable and wireless communications are based on RF modulated signals in the 100 MHz (cable) or 1 to 5 GHz band (wireless). To construct chipsets for these architectures, still more expensive technologies like SiGe are preferred [Cloetens, 2001].

The deployment of wireless broadband access has the advantage that no extra cabling is necessary and maintenance cost will be lower. But the set-up cost of a large number of base-stations cannot be recovered yet by the present market situation. Also the energy that needs to be send in order to reach a sufficient bandwidth, will be rather high. So wireless is for the moment only supposed to break through in the last yard access technologies like Wireless LAN and Bluetooth.

As a conclusion, it can be stated that although technology seems to be ready to connect everybody by an optical fibre to the home, the market clearly is not. Since no valuable application justifies the enormous investments to build up this network, fibre will be left in the fridge for several more years. The two remaining competing technologies : coax and DSL have both their merits since they reuse the existing connecting network. In this, DSL has the advantage of a higher coverage, certainly in business areas. The best summary of this section is given by Ray Smith of Bell Atlantic : " ADSL is an interim technology, for the next forty years. "

2.   The Channel

2.1   The Distribution Network

Distribution cables contain 25 to 1000 pairs. For residential and small business area, the distribution cables lead to the drop wire that serves each customer. The distribution cable connects to the drop wires. Typical drop wires contain two or three pairs of 22 AWG, although larger numbers are found in some areas. Many of the drop wires installed prior to 1992 were not twisted.

The feeder and distribution cables are bundled into binder groups of 25, 50 or 100 pairs. The pairs within a binder group remain adjacent to each other for the length of the cable. As a result, the crosstalk of pairs within a binder group is somewhat greater than crosstalk between pairs in different binder groups. Despite the administrative complexities, telephone companies will sometimes segregate certain services (such as T1 carrier) into separate binder groups.

Cables connecting to a CO nay have up to 10,000 pairs. As one follows the cable plant from the CO to the customers site, the cables branch. As a result, fewer customer lines are accessible at points closer to the customer site. The number of wire pairs per cable becomes progressively smaller at successive splice points approaching the customer site. Feeder and distribution cable pair counts were traditionally sized to meet the service demand forecast for 20 years from the construction date. More recently, cable designs have been based on a shorter service capacity life. Also, the demand for more than one line per living unit has grown far beyond what was expected. As a result, there is a strong need to conserve wire pairs. This is addressed by ADSL's ability to convey POTS and data on one wire pair.

2.2   Channel Properties

Twisted-pair telephone wires can be well modelled for transmission by their two-port model for frequencies up to 30 MHz. The model can be used to calculate the voltage and currents at the origin of a line with length d, when a

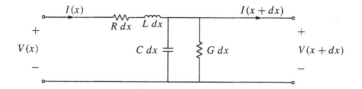

Figure 2.8: RLC transmission line model per unit length

signal is applied at the other end of the line.

$$\begin{bmatrix} V(x=0) \\ I(x=0) \end{bmatrix} = \begin{bmatrix} \cosh{(\overline{\gamma}d)} & Z_0 \sinh{(\overline{\gamma}d)} \\ \frac{1}{Z_0} \sinh{(\overline{\gamma}d)} & \cosh{(\overline{\gamma}d)} \end{bmatrix} \begin{bmatrix} V(x=d) \\ I(x=d) \end{bmatrix} \quad (2.4)$$

$$\begin{bmatrix} V(x=0) \\ I(x=0) \end{bmatrix} = \Phi \begin{bmatrix} V(x=d) \\ I(x=d) \end{bmatrix} \quad (2.5)$$

in which Z_0 denotes the characteristic impedance and γ the complex attenuation constant. The big advantage of using this two-port model is the ease by which the overall two-port model can be calculated if several ports are connected in series. For instance if wires with different gauges are used to connect the CO with the CPE, the overall two-port model will become :

$$\begin{bmatrix} V(x=0) \\ I(x=0) \end{bmatrix} = \Phi_1 \times \Phi_2 \times \Phi_3 \begin{bmatrix} V(x=d_1+d_2+d_3) \\ I(x=d_1+d_2+d_3) \end{bmatrix} \quad (2.6)$$

For a line that is terminated with an impedance Z_L, the line transfer function can be calculated from (2.4) :

$$T(f) = \frac{Z_L}{Z_L \cosh{(\overline{\gamma}d)} + Z_0 \sinh{(\overline{\gamma}d)}} \quad (2.7)$$

The input impedance Z_{in} can also be directly derived from the two-port representation :

$$Z_{in} = Z_0 \frac{Z_L + Z_0 \tanh{(\overline{\gamma}d)}}{Z_0 + Z_L \tanh{(\overline{\gamma}d)}} \quad (2.8)$$

The characteristic impedance and complex attenuation factor can be derived from the transmission line model per unit length, given in figure 2.8. The R,L,C and G values of real cables however don't follow smooth curves. Other models are needed to fit the characteristics of practically used lines. The models are :

$$R(f) = \sqrt[4]{r_{0c}^4 + a_c f^2} \quad (2.9)$$

In which r_{0c} denotes the copper Direct Current (DC) resistance, while a_c is a constant that models the increase in resistance with frequency due to the skin effect. The inductance is modelled by :

$$L(f) = \frac{l_0 + l_\infty \left(\frac{f}{f_m}\right)^b}{1 + \left(\frac{f}{f_m}\right)^b} \tag{2.10}$$

Where l_0 and l_∞ are the low-frequent and high-frequent inductance, respectively f_m is the transition frequency. The transition below low and high frequencies is characterised by the b parameter. The capacitance in function of frequency is fitted by :

$$C(f) = c_\infty + c_0 f^{-c_e} \tag{2.11}$$

Again c_∞ denotes the contact capacitance, while c_0 and c_e are constants to fit the measurements. An almost similar fit is used for the unit length conductance :

$$G(f) = g_0 f^{g_e} \tag{2.12}$$

As an illustration of a typical wire-pair transfer function, the bode plot of a

Table 2.3: Model parameters for common wire types

		26AWG[a]	24AWG[b]	0.5 DWUG[c]	DW10[d]	DW8[e]
r_{0c}	[Ω/km]	286.18	174.56	179.0	180.93	41.16
a_c	[ ]	0.1477	0.0531	0.0359	0.0497	0.0012
l_0	[μH/km]	675.37	617.30	695	728.87	1000
l_∞	[μH/km]	488.95	478.97	585	543.43	911
f_m	[kHz]	806	554	1000	719	175
b	[ ]	0.93	1.15	1.2	0.76	1.20
g_0	[nS/km]	43	$0.23 \, 10^{-3}$	0.5	89	53
g_e	[ ]	0.7	1.38	1.033	0.86	0.88
c_∞	[nF/km]	49	50	55	51	23
c_0	[nF/km]	0	0	1	63.8	32
c_e	[ ]	0	0	-0.1	0.12	0.11
R[f]	[Ω/km]	457	345	316	341	132
L[f]	[μH/km]	603	552	662	649	931
C[f]	[nF/km]	49	50	59	65	30
G[f]	[μS/km]	420	17.2	385	6730	5490

[a]Twisted pair wire with diameter of 0.405 mm
[b]Twisted pair wire with diameter of 0.51 mm
[c]British Telecom 0.5 mm distribution cable
[d]British Reinforced 0.5 mm copper PVC insulated drop wire
[e]1.14 mm flat untwisted pair
[f]@ 500 kHz

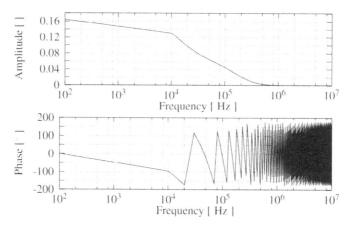

Figure 2.9: Line transfer function of a 12 kft 24AWG cable, terminated by a 125 Ω termination resistance, plotted with a linear vertical axis to stress the small passband

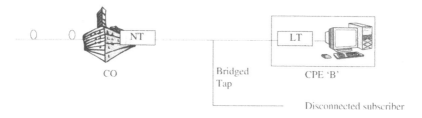

Figure 2.10: Schematic representation of a bridged tap

12 kft 24AWG cable is given in figure 2.9. The line is terminated by a 125 Ω termination resistor. The line transfer function, which is the ratio of the output voltage on the terminating load resistance over the a applied input voltage, is obtained by calculating the two-port model using the parameters of table 2.3.

2.3 Cable Impairments

2.3.1 Bridged Taps

In some countries, there is a common practice of splicing a branching connection (called a bridged tap) onto a cable as in figure 2.10. Thus, a bridged tap is a length of wire pair that is connected to a loop at one end and is unterminated at the other end. This is mostly due to a subscriber that is disconnected from service. Normally, the wire should be disconnected up to the trunk, but mostly it seemed more economical at those times to just disconnect the wire at the customers side. Approximately 80% of loops in the United States have

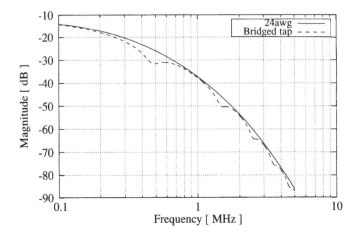

Figure 2.11: The effect of bridged taps on the line characteristic

bridged taps; sometimes several bridged taps exist on a loop. Bridged taps may
be located near either end or at an intermediate point. Most countries in Eu-
rope claim to have no bridged taps, but there have been reports of exceptions.
The reflections of signals from the unterminated bridged taps result in signal
loss and distortion. The two port model of a bridged tap with length d_t and
characteristic impedance Z_{0t} is :

$$\Phi_t = \begin{bmatrix} 1 & 0 \\ \frac{1}{Z_{0t}} \tanh(\overline{\gamma} d_t) & 1 \end{bmatrix} \tag{2.13}$$

The effect of one bridged tap is depicted in figure 2.11. The transfer function
of a 12 kft 24AWG line is given next to the same line, but with a 100 m long
24AWG bridged tap at 200 m from the end of the line.

2.3.2   Cross-Talk

Since xDSL uses unshielded wire pairs as a transmission media, the lines
couple electromagnetically in the bundle. Two types of crosstalk are dis-
tinguished Near-End Crosstalk (NEXT) and Far-End Crosstalk (FEXT). Figure 2.12 shows both crosstalk mechanisms in a schematic fashion.

NEXT occurs between a transceiver and a receiver at the same premises.
Since the crosstalk signal does not pass through the line, it is much bigger than the attenuated receive signal. However, it is theoretical possible
to employ echo cancellation between transmitters of the same bundle at the
CO-side. Due to the enormous complexity, this is not implemented yet.

FEXT runs from one side's transceiver to a receiver at the other side. FEXT
is thus attenuated by the channel like the wanted receive signal.

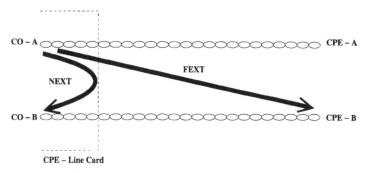

Figure 2.12: Schematic overview of NEXT and FEXT

Table 2.4: Amateur Radio Bands in the DSL range

lowest frequency [MHz]	highest frequency [MHz]
1.81	2.0
3.5	4.0
7.0	7.1
10.1	10.15
14.0	14.35
18.068	18.168
21	21.45
24.89	24.99
28.0	29.7

Note that for bundles that still carry ISDN, the crosstalk requirements also take those specifications into account.

2.3.3 Ingress and Egress

Another issue related with the unshielded nature of the telephone wire is radio noise. It is the remnant of wireless transmission signals on phone lines, particularly Amplitude modulation (AM) radio broadcasts and amateur (HAM) operator transmission.

RF signals impinge on twisted-wire lines, especially aerial lines. Phone lines, being made of copper, make relatively good antennae leading to an induced common mode voltage on the twisted pair. Well-balanced twisted pairs thus should see a significant reduction in differential RF signals on the pair. However, balance decreases with increasing frequency, and so at frequencies of DSLs from 560 kHz to 30 MHz, DSL systems can overlap radio bands and will receive some level of RF noise along with the differential DSL signals on the same phone lines. This type of DSL noise is known as RF ingress.

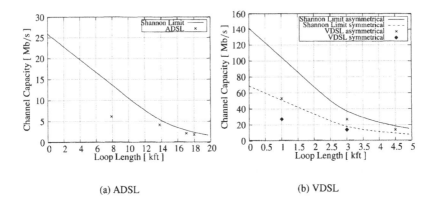

(a) ADSL   (b) VDSL

Figure 2.13: Shannon limitation of a telephone cable compared with ADSL and VDSL specifications

The possible HAM radio bands are shown in table 2.4. These bands overlap the transmission band of VDSL but do not directly intervene the lower frequent ADSL-bands. Thus ingress is a major problem only for VDSL.

The complementary of RF ingress, being the emission of signals in the HAM radio bands by the DSL transmitter is an issue that affects all DSL types. Since distortion up-converts signals in higher frequency bands the out-of-band specifications are also limited by egress. The standard imposes a maximum allowed power density of -80 dBm/Hz in the standardised HAM radio bands [Jacobsen, 1999].

3. Modulation Techniques

3.1 The Shannon Limit

Claude Shannon calculated a theoretical limit for the capacity of a channel that is limited by white Gaussian noise. This limit can be regarded as the maximal achievable bit-rate C_{TP} through a channel. It is given by :

$$C_{TP} = BW \, \log_2 \left(1 + \frac{S}{N} \right) \qquad (2.14)$$

in which BW is the channel's bandwidth and S/N is the signal to noise ratio in the channel. For a lossy channel this limitation is calculated by :

$$C_{TP} = \int_{f_{min}}^{f_{max}} \log_2 \left(1 + \frac{S}{N} \right) df \qquad (2.15)$$

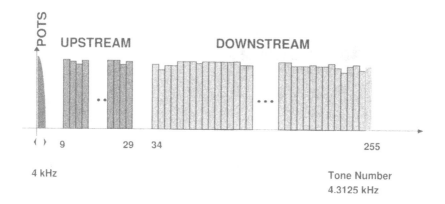

Figure 2.14: Spectrum of a DMT-modulated ADSL-signal

This formula is calculated for the ADSL downstream (DS) band and the symmetrical and asymmetrical ETSI VDSL DS band plan. The results of these calculations are given in figure 2.13 together with the reported maximal bit-rates [Cornil et al., 1999a]. For long loops the ADSL and VDSL bit-rates approach the maximal achievable Shannon limit.

3.2 DMT-modulation

3.2.1 Basic Properties

xDSL uses DMT-modulation to reach these close-to-Shannon bit-rates. The spectrum of an DMT modulated signal is given in figure 2.14. The basic idea of DMT-modulation is :

- split the bandwidth into several discrete channels.

- for every channel the SNR is measured.

- every channel is modulated by n-bit Quadrature Amplitude Modulated (QAM)-modulation. The constellation size is determined by the measured SNR and in accordance with Shannon's relation (2.14). This method is called the water-filling method [Kalet, 1989].

By discretising the twisted pair bandwidth in several channels, a very good approximation of the integral form of Shannon's limit (2.15) is obtained. Of course, there are other possibilities to obtain a good approximation, but the main advantage of DMT is that it is easy to implement this modulation in the digital domain by a mere Inverse Fast Fourier Transform (IFFT).

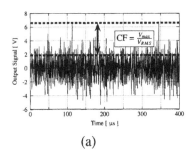
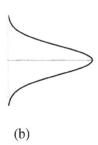

(a)   (b)

Figure 2.15: Time domain representation of a DMT-modulated signal (a) and amplitude distribution (b).

Figure 2.14 shows a sketch of the spectrum used for ADSL communication. The Discrete Multi-Tone (DMT) modulation in this case consists of 22 carriers containing the upstream (US) information and 221 containing the downstream (DS) information.

3.2.2   Time Domain Representation - The Crest Factor

The time domain representation of the signal is given in figure 2.15. Since the spectrum is a flat spectrum, the time domain representation will have a noise-like nature. An important property of the signal is clearly shown. When the phases of several carriers align in constructive interference, a voltage spike occurs in the signal. A measure for the height of this spike is the CF. It is defined as the maximum voltage over the root mean square (rms) voltage :

$$CF = \frac{V_{max}}{V_{rms}} \qquad (2.16)$$

Some other publications tend to use the term Peak-to-Average-Ratio (PAR). It is defined as the peak power over the rms-power of the signal.

The time-domain ADSL signal can be represented by :

$$x(t) = \sum_{n=34}^{255} a_n s(t - nT) \cos(2\pi n f_{it}) + \sum_{n=34}^{255} b_n s(t - nT) \sin(2\pi n f_{it}) \quad (2.17)$$

In this equation, a_n and b_n represent the constellation point of the m-bit[8] QAM modulation, $s(t)$ is an envelope function used to normalise the average energy per channel. The carrier spacing f_i is set to 4.3125 kHz for the ADSL

---

[8]The constellation size m is on itself function of the carrier number n, so m(n)-bit QAM would be a better notation

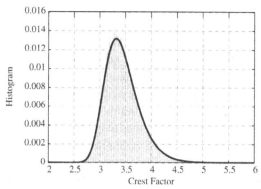

Figure 2.16: Distribution function for the crest factor of an upstream ADSL symbol (solid line) compared with a monte carlo simulation (grey histogram).

system. An ADSL symbol has thus a length of $1/f_i$. During this time the constellation points stay fixed.

Since the constellation points can be regarded as random variables for normal data communication and the number of carriers is sufficiently large for the central limit theorem to hold [MestDagh et al., 1993], the amplitude distribution f_A of the DMT-signal $x(t)$ can be approximated as being Gaussian (2.18).

$$f_A = \frac{1}{\sigma\sqrt{2\pi}} \exp\left(-\frac{A^2}{2\sigma^2}\right) \qquad (2.18)$$

with σ representing the rms-voltage of the signal. This is graphically represented in figure 2.15. One can conclude that one has a finite probability that a symbol contains a high voltage peak. Since the CF is an important parameter to determine the specifications of the AFE, the maximal CF of a DMT-symbol needs to be calculated. This can be done by taking the Gaussian distribution assumption for the voltage distribution. The distribution function for the CF in an average DMT generated symbol with N carriers is then an extreme value distribution :

$$f_{CF} = \frac{2\sqrt{2}}{\sigma\sqrt{\pi}} N \left(\mathbf{erf}\left(\frac{\sqrt{2}CF}{2\sigma}\right)\right)^{2N-1} \exp\left(-\frac{CF^2}{2\sigma^2}\right) \qquad (2.19)$$

By filling in the parameters of an ADSL-symbol, the crest factor distribution for an ADSL DS signal can be calculated. The result is plotted in figure 2.16 and compared with the histogram obtained from a Monte Carlo simulation. For this, the histogram of the crest factor of 50000 randomly generated 4QAM modulated ADSL downstream signals has been calculated. This has a slow tail when going to high crest factors. Unlike the ideal Fisher-Tippet distribution, the distribution of the crest factors of an ADSL-downstream signal has a maximum of 22.6, when all the carriers are in phase.

If the channel is assumed to be ideal and an ideal transmission is required ; the system needs to be designed so it an cope with signals that are 22.6 times larger than their rms value. If not an amount of symbols will be corrupted, as can be predicted from figure 2.16. The achievable Bit-Error Rate (BER) will thus be directly coupled with the maximum CF a system can deal with.

In practical designs the crest factor has to be limited. The clipping of the signal will lower the quantisation noise of the used Analog-to-Digital Converter (ADC) and the Digital-to-Analog Converter (DAC) and their power consumption, but will introduce clipping noise [MestDagh et al., 1993]. In practical ADSL systems the CF is limited to a factor of 15 dB or 5.6, which is still a reasonably large number.

3.2.3   Clipping Noise

Due to the nature of the time-domain signal, a non-linearity will generate a rise in the noise-floor of the complete signal. This 'distortion-noise' can be calculated with the cross-correlation of the original signal with the output of the non-linear characteristic $g(x)$ [Tellado-Mourelo, 1999].

$$\sigma_d^2 = \int_{-\infty}^{\infty} (x - g(x))^2 \exp\left(-\frac{x^2}{2\sigma^2}\right) dx \qquad (2.20)$$

Clipping noise can be regarded as a special case of (2.20).

$$\sigma_d^2 = \int_{A}^{\infty} (x - A)^2 \exp\left(-\frac{x^2}{2\sigma^2}\right) dx \qquad (2.21)$$

Due to the nature of the DMT-signal, a direct relation between Signal-to-Noise Ratio (SNR) and MTPR cannot be calculated without information on the nature of the non-linearity.

This distortion-noise will not only be the limiting factor for the in-band bit-rate, but the out-of-band distortion-noise will jam the upstream signal coming from the CPE-side. This distorted echo-signal is very hard to compensate in the digital domain.

3.3   DMT Specifications

3.3.1   Missing Tone Power Ratio and Missing Band Depth

As been shown in section 2.3.2.3, a non-linearity will express itself as an increase of the noise-level. Since the bit rates of a carrier is limited by the SNR of the channel in a carrier bin, the maximal achievable bit rate is directly dependent on the linearity of the system. However, due to the nature of the signal as shown in (2.20), the bit-error rate cannot be coupled directly to standard distortion specifications like HD3, IM3, etc.. This can be understood as follows : since the bit-error rate of one carrier is determined by the instantaneous[9] signal-to-noise ratio in that channel. If only distortion is taken into account, the noise-level is formed by summing all distortion contributions from all other carriers. To do this summation not only the magnitude of the distortion contributions needs to be known, but also the phase-shift of every component. Since phase information is not included in the standard distortion specifications, and thus the complete characteristic of the non-linearity needs to be known, in principle, an infinite number of standard distortion specifications (HD2, HD3, HD4, ...) needs to be taken into account to determine the ADSL specifications. As thus the nature of the non-linearity needs to be known in order to give an estimate for the achievable bit rates, other specification figures were devised to test the correctness of a proposed xDSL solution.

The main test for an ADSL system is the Missing Tone Power Ratio or MTPR. The MTPR is measured by applying a signal where all carriers are modulated by a DC signal. The spectrum will thus consist of discrete frequency peaks. Some tones are not activated. These will act as antenna-tones. Due to a non-linearity all distortion components will add up in these gaps. The ratio of the Power Spectral Density (PSD) of the activated tones over the power found in the antenna-tones is called the MTPR. This is illustrated in figure 2.17. The MTPR figure is the average of the measurements with several symbols. Since a DC-signal is transmitted, the central limit theorem is not applicable and the distortion noise will thus be 'coloured' noise and the obtained results could be manipulated by selecting a favourable input symbol. By averaging several measurements this problem is elevated. Also together with an MTPR measurement the CF of the symbol needs to be reported.

For VDSL, the MTPR measurement does not make sense no longer, due to the high number of tones. To cope with this objection, the Missing Band Depth (MBD) is defined. Since most VDSL systems are organised in two or more bands[10], the US or DS bands are activated. Again due to distortion, the tones in the disactivated bands will be filled with spurious peaks. The ratio between the PSD of the activated band and the height of the spurious tones, averaged out over 10 spurious tones is the Missing Band Depth (MBD).

---

[9]With instantaneous, the integrated SNR over one symbol length is meant. The non-linearity is approximated to be memory-less.

[10]The Zipper VDSL system is a system that is not organised in bands. The carriers are alternating dedicated to carry upstream/downstream signals. Theoretically this will lead to the highest possible perfectly symmetrical bit rate, but it is too demanding to build an AFE for it due to the heavy cross-talk

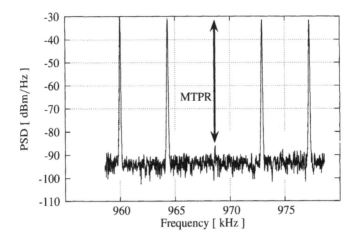

Figure 2.17: Illustration of the definition of the MTPR

3.3.2   Spectral Masks

The other specifications of an xDSL are summarised in the following spectral masks. Figure 2.18 shows the spectral mask as stated by the T1 specifications [Wang, 2001]. To allow POTS compatibility the generated spurious peaks
should be below -97 dBm/Hz in the POTS-band. Since NEXT is a major issue in ADSL-systems the PSD is limited to -72 dBm/Hz peak at 80 kHz in
the US band in order to lower the received NEXT power. The nominal power
PSD in the DS is -40 dBm/Hz with a margin of 3 dB. Above 3 MHz, the
peaks should be below -90 dBm/Hz, and the integrated power in a 1 MHz
sliding window should be below -50 dBm. The latter is shown by the dashed
line in figure 2.18.

The VDSL specifications are shown in figure 2.19. Both the Fibre To The
Exchange (FTTEx), i.e. the VDSL channel is deployed from the CO directly
to the CPE, and Fibre To The Cabinet (FTTCab), i.e. the local cabinet case
holds the network termination, deployment schemes are sketched. In the FT-
TEx case, the line driver also needs to drive the ADSL power. In the ADSL DS
the transmitted power is thus limited to -40 dBm/Hz. The FTTCab deployment scenario only transmits local VDSL signals, and for power dissipation
reasons, the complete band is limited to -60 dBm/Hz. The maximum transmitted power in a VDSL system is 14.5 dBm. Out of band specifications limit
the PSD to -100 dBm/Hz outside the transition bands of 175 kHz. At the be-
gin of a transition band, the transmit power density is limited to -80 dBm/Hz.

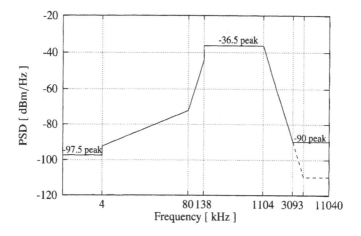

Figure 2.18: Downstream ADSL spectral masks according the T1 specifications

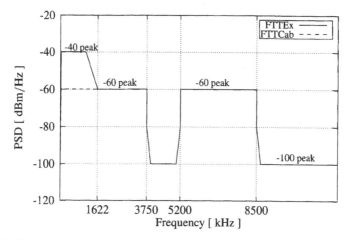

Figure 2.19: VDSL FTTEx and FTTCab spectral masks according the T1 specifications

4. Driving the Line

4.1 The Challenge

In table 2.5 the most important properties for line driver design of ADSL and VDSL are summarised. To compare these specifications with older technologies, one can clearly notice the enormous shift in line driver requirements for the two digital subscriber modem types. In less than one decade of xDSL research activities, an improvement of over 3 decades in bit-rate has been reached. This however at a drastic cost in the AFE design specification. The

Table 2.5: Summary of the most important xDSL requirements for line driver design. The V.34 standard is added in grey for comparison reasons.

	Distortion	*Bandwidth*	*Output Power*
V.34	THD <-70 dB	4.96 kHz	0 dBm
ADSL(-Lite) US[‡]		103.5 kHz	13 dBm
ADSL-Lite DS[†]	MTPR > 34 dB	418.3 kHz	16.3 dBm
ADSL DS[†]	MTPR > 55 dB	970.3 kHz	20 dBm
VDSL DS[†]	MBD > 63 dB	8.5 MHz	14.5 dBm

[†] DS = downstream, [‡] US = upstream

linearity specifications are not significantly relaxed, but the bandwidth and necessary output power are increased towards the current technology limits.

For the choice of a process technology, one has to take into account that the twisted pair is also shared with the Plain Old Telephony System (POTS). On the copper wire high voltages occur due to the POTS system. A line interface that is able to handle POTS and xDSL on the same die, has to take these high voltages into account [Zojer et al., 1997, Benton et al., 2001]. Furthermore a POTS line card has standard a 48 V supply. The use of every other voltages will require an extra DC-DC converters. Their efficiency adds up in the total power budget.

As a summary, due to the high performance of an xDSL system, the line driver has to:

- Drive a relative large power to the line

- with a high linearity

- in a high bandwidth.

- The power dissipation needs to be minimal to allow maximal install-able lines at the CO-side

- Out-of-band specifications are very stringent (<-100 dBm/Hz)

- Deal with high voltages of the POTS signals

4.2   Line Termination and The Hybrid

The hybrid is the final building block between the line driver and the line. Its function is twofold :

- The hybrid needs to terminatetermination the line properly in order to suppress reflections.

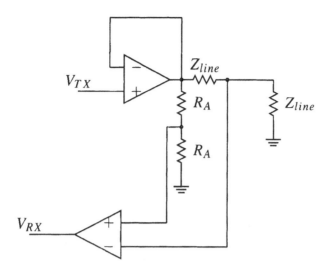

Figure 2.20: Resistive termination network and hybrid circuitry

- Since the xDSL communication system is a full-duplex system, the receive signal needs to be split off. The hybrid is a three port function which ideally transfers the transmit signal directly to the line but blocks it to the receive input. The receive signal from the line should be transferred to the inputs of the receive path without too much degradation.

Figure 2.20 shows a possible implementation of this function. The line is resistively terminated. If the termination is perfect, the receive signal on top of half the transmit signal is found at the line input. By dividing the output of the line driver by two and subtracting both signals, the receive signal can be sensed. This circuit, however, has several disadvantages :

- The resistance is a rather bad approximation of the line's impedance as shown in section 2.2. The hybrid suppression is for xDSL thus maximal -15 dB. Attempts were made to improve the termination [Laaser et al., 2001], but this at the cost of more external components.

- The hybrid, due to its high matching properties is mostly made of external components. So, it consumes board space and increases the total cost.

- Half the output power of the line driver is dissipated in the termination resistance. Maximum efficiency is thus limited to 50% from the very start.

- The output swing is halved. Therefor a higher transformer turn ratio is required or a higher voltage Integrated Circuits (IC) technology is required.

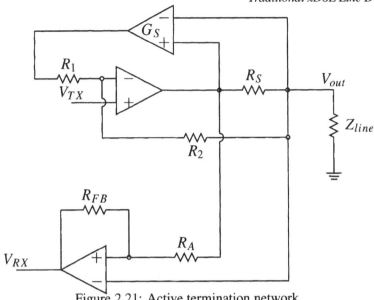

Figure 2.21: Active termination network

To increase the efficiency, active line termination is used. Figure 2.21 shows a possible implementation [Bicakci et al., 2003]. The input-output relation is given by:

$$V_{out} = \frac{1 + \frac{R_1}{R_2}}{R_L + G_S R_S \frac{R_1}{R_2}} V_{in} \tag{2.22}$$

The output impedance seen from the line is given by :

$$Z_{out} = G_S R_S \frac{R_1}{R_2} \tag{2.23}$$

In this way the series resistance R_S can be made G_S times smaller than in the resistive termination case. It can, however, not be made too small for noise reasons. Practical implementations lead to a value for R_S which is approximately 10% of the line resistance.

The receive signal (V_{RX}) can then be extracted if the following balance equation holds :

$$\frac{R_{FB}}{R_A} = \frac{Z_{line}}{R_S} \tag{2.24}$$

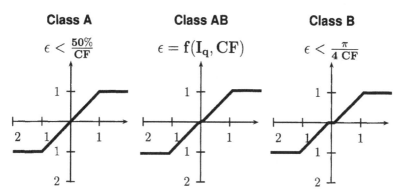

Figure 2.22: Input-output characteristics for class A (left), class AB (middle) and class B (right) operation.

5. Solutions for xDSL Line Drivers
5.1 Class AB

Due to the high bandwidth and the high linearity specifications of an xDSL system, a linear power amplifier was the natural choice and all initial designs were targeted towards class AB line drivers. In this type of amplifier the active elements will conduct the output currents while there is an output voltage over these elements. The efficiency of this type of amplifiers will thus be very low.

5.1.1 Class AB operation

The class AB operation of a power amplifier is an intermediate between class A and class B. Class A is the most linear type of power amplifier. This is due to the fact that the active element (output driving transistor) is never turned off. This comes with a significant heat-dissipation and thus very low efficiency. Class A operation is depicted at the left of figure 2.22. Above the input-output relationship in figure 2.22, the asymptotic formulas for the efficiency to drive a sinusoidal signal is depicted. For a crest factor of 5.6 and a rail-to-rail output driving stage (which is in practice impossible) the maximum efficiency for a sinusoidal signal will be lower than 9%. The theoretical minimal power dissipation of a class A power amplifier for ADSL is more than 2 W, so the class A operation will be completely unacceptable for an xDSL system.

The class B operation gives a higher efficiency by taking two active elements in a push-pull configuration. Power is saved since the pull transistor can be turned off for the falling edge and vice-versa. For a DMT-signal the efficiency can be calculated as :

$$\epsilon = \frac{\sqrt{2\pi}}{2} \frac{V_{rms}}{V_{DD}} = \frac{\sqrt{2\pi}}{2} \frac{1}{CF} \frac{V_{swing}}{V_{DD}} \tag{2.25}$$

If we assume a rail-to-rail output swing a maximum efficiency of 22% can be reached. For a resistive terminated ADSL-line, the theoretical class B line driver thus dissipates 710 mW.

However due to the threshold voltage (V_T) of a transistor, the two active elements of the push-pull stage are both turned off around the zero-crossing. Therefor the input-output characteristic of an ideal class B power amplifier shows a dead zone. Considering the general distortion-noise calculation (2.20), the distortion will be active in the region where the signal resides most of the time. The distortion-noise associated with crossover-distortion can be calculated as follows [Tellado-Mourelo, 1999]:

$$\sigma_d^2 = \int\limits_{-V_T}^{V_T} x^2 \exp\left(-\frac{x^2}{2\sigma^2}\right) dx \qquad (2.26)$$

Since in the dead-zone the gain of the power amplifier is zero, this type of distortion cannot be lowered by applying feedback. To cope with this , a biasing scheme between the two active elements is added. In this way the elements are never turned off and the dead-zone is transformed into a zone with lower but non-zero gain. Feedback can restore linearity to allowable levels. At the zero-crossing, a small current, the quiescent current, flows through both the active elements. Equation 2.26 has to be altered thus with the quiescent current linearised input-output characteristic $g(x, I_q)$.

$$\sigma_d^2 = \int\limits_{-V_T}^{V_T} (x - g(x, I_q))^2 \exp\left(-\frac{x^2}{2\sigma^2}\right) dx \qquad (2.27)$$

The quiescent current (I_q) has to be accurately controlled, since an I_q which is too low, will generate too much distortion. The higher the I_q, the closer the amplifier will get to class A operation, thus the lower the efficiency. Moreover, care has to be taken to prevent too much power consumption when the amplifier is in the off-state. For the calculation of the overall efficiency the power-consumption of the quiescent current control circuitry has to be added (P_{qcc}) as well.

$$\epsilon = \frac{V_{rms}^2 \sqrt{\pi}}{V_{DD}(V_{rms}\sqrt{2} + I_q\sqrt{\pi}) + P_{qcc}\sqrt{\pi}} \qquad (2.28)$$

Assuming a rail-to-rail output and a fixed CF for DMT-modulation, the efficiency for a class AB line driver can be written as

$$\epsilon = \frac{V_{DD}^2 \sqrt{\pi}}{CF\, V_{DD}(V_{DD}\sqrt{2} + CF\, I_q\sqrt{\pi}) + CF\, P_{qcc}\sqrt{\pi}} \qquad (2.29)$$

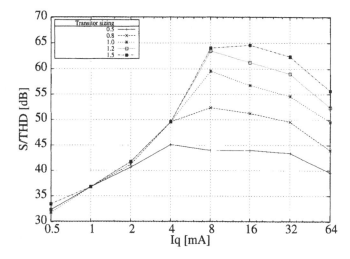

Figure 2.23: Signal to total distortion ratio versus quiescent current [Casier et al., 1998b].

The efficiency for a class AB line driver increases with increasing supply voltage [Benton et al., 2001], ultimately approaching the class B limit. This however requires a more expensive, high frequent, high voltage technology. The class AB operation is currently the favourite xDSL line driver [Benton et al., 2001, Cresi et al., 2001, Ingels et al., 2002, Sabouti and Shariatdoust, 2002], for its high linearity. By using a high voltage technology and active back termination its power consumption can be limited to 750 mW [Moons, 2003] which is very close to the minimal theoretical power consumption. Class AB line drivers thus have a maximum efficiency of 13%.

5.1.2 Quiescent current control

When designing class AB line drivers, the quiescent current control block is the most critical block for the performance of a class AB line driver.

Figure 2.23 shows the dependence of the quiescent current and the transistor sizing on the signal to total harmonic distortion ratio, taken from simulation results presented in [Casier et al., 1998b]. In this simulation three distinct regions can be observed. In the low I_q regions, the cross-over distortion is dominant in the signal to distortion ratio. The obtained Signal-to-Noise-and-Distortion (SNDR) levels are independent from the relative transistor sizing. In the middle region, the class AB line driver is in its optimal performance region. SNDR becomes highly dependent on the transistor sizing since the maximal driving capacity defines the output signal level. In the high I_q region, the output transistors are put more quickly in the linear operating region. The distortion level rises due to signal clipping.

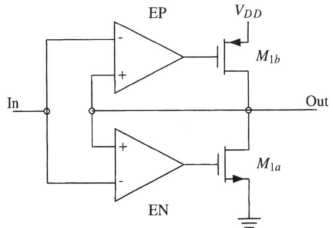

Figure 2.24: Line driver with included error-amplifiers to lower cross-over distortion.

A first approach to lower the influences of the cross-over distortion has been presented in [Khorramabadi, 1992]. Two error amplifiers are placed in connection with a pseudo push-pull connected pair to lower the cross-over distortion. This is schematically presented in figure 2.24. The increase in SNDR can be easily calculated (2.30).

$$HD_{CL} = \frac{HD_0}{A_{preamp} A_{EP|N} \, gm_{M1a|b} \, R_{load}} \qquad (2.30)$$

So the initial amount of output-related cross-over distortion HD_0 is lowered by the loop gain containing the error amplifier gain and the gain of the pre-amplifier. This reduction however will be limited by the offset voltage of the error amplifiers. The estimated maximum variation on the quiescent current ΔI_q can be calculated as

$$\frac{\Delta I_q}{I_q} = \frac{2V_{offset} A_{EP}}{(V_{gs} - V_T)_{M1a|b}} \qquad (2.31)$$

Together with the results from figure 2.23 and using a technology with reasonable matching, the maximum error amplification A_{EP} is around 8.

To overcome the problem of a random offset in the error-amplifiers, a quiescent current control scheme has to be adopted. Figure 2.25 shows schematically an approach adopted in [Casier et al., 1998b]. The output transistors are decoupled in order to measure the quiescent current through the transistors $M_{1a|b}$. Since the output transistors are quite large this measurement will match the quiescent current through the actual output transistors $M_{2a|b}$. The measured I_q will then be compared with a reference current I_{ref}. The quiescent current control circuit used in [Casier et al., 1998b] consist of two Schmitt

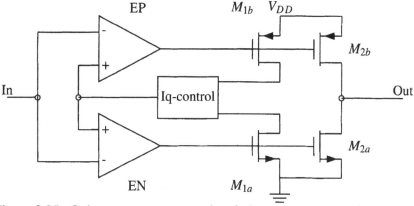

Figure 2.25: Quiescent current control technique to overcome limited error-amplification due to random offset.

triggers (one for the output PMOS, one for the output NMOS) which trigger a charge pump. This charge pump together with the dump capacitor form a filter with a very small time constant in order to only get the DC quiescent current. The obtained error signal is added at the inputs of the error amplifiers to counteract the random offset. An SNDR of 62dB for a 100 mW signal has been reported using this technique. The reference current can be generated on-chip by a bandgap reference. This however limits the flexibility of the control loop.

Controllability has been gained in [Ingels et al., 2002] by replacing the analog quiescent control loop by a digital one. The cost to be paid is a loss in efficiency by the integration of a measurement ADC and a controlling DAC.

5.1.3   Final Remarks on class AB

The class AB line driver is still the most widely used line driver for its high intrinsic linearity and its well known behaviour. New circuit techniques allow designers to construct quiescent current control loops which have a lower power consumption but provide higher linearities over a wider mismatch range.

However due to its very low power efficiency, the class AB line driver is the power bottleneck for xDSL systems and will become unusable in future CO applications.

5.2   Class G/H

The class G line driver is a logical solution for the efficiency problem related with class AB line drivers. The efficiency of a class AB line driver is inversely proportional with the ratio of the rms voltage V_{rms} and the supply voltage V_{DD} (2.28). In a class G design, multiple supplies are connected to a class AB line driver. In this section we will further focus on the case were two supplies are

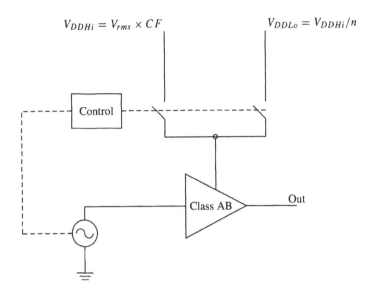

Figure 2.26: Principle scheme of a 2 supply class G power amplifier.

used : one scaled supply for the main signal V_{DD}/n and one higher for the peak amplitudes V_{DD}. Figure 2.26 shows the topology which is mostly used nowadays [Pierdomenico et al., 2002]. The efficiency for an ideal two-supply class G amplifier can be calculated as follows :

$$\epsilon = \frac{n \, V_{rms} \sqrt{\pi}}{V_{DD}\sqrt{2} \left(\left(1 - \exp\left(\frac{-V_{DD}^2}{2n^2 V_{rms}^2}\right)\right) + n \exp\left(\frac{-V_{DD}^2}{2n^2 V_{rms}^2}\right)\right)} \tag{2.32}$$

Figure 2.27 shows the obtained efficiency for different values of $n = \frac{V_{DDHi}}{V_{DDLo}}$ with V_{DDHi} obtained from the crest factor constraints described in section section 2.3.2.2. Values below n=1 create supply voltages that are too high for the crest factor constraint, resulting in an extremely low efficiency. At n=1 the class AB efficiency is obtained. The efficiency approaches a maximum around n=3. This maximum is below 50% for a 2 supply class G amplifier. For higher values of n more amplitudes are amplified using the high supply. Ultimately the class AB efficiency is reached.

These figures are however to optimistic and state-of-the-art class G xDSL power amplifiers show efficiencies close to their high-voltage class AB counterparts [Pierdomenico et al., 2002]. This is mostly due to :

- The matching between the delays of the forward signal path through the class AB power amplifier and the supply steering path (dashed line in figure 2.26) is of the utmost importance. Any mismatch will generate clipping noise (2.21) like a DMT-signal clipped by $V_{DDLO} = V_{DD}/n$. Since this

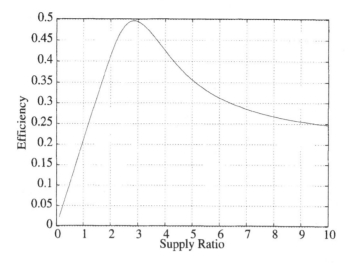

Figure 2.27: Efficiency versus V_{DDHi}/V_{DDLow} ratio for an ADSL input signal.

timing delay constraint is relative to the bandwidth of the signal, it will be even more important towards the faster xDSL types. Therefor class G will only be feasible when the detection circuitry, the class AB power amplifier and the supply switches are integrated on the same die. To relax the matching constraint, a larger envelope around the voltage peak has to be generated. This comes at the cost of a more elaborate analogue signal processing and less efficiency due to a suboptimal supply switching scheme.

- The generation of the switching signal is very critical. The detection of the input signal at the trigger level for the switches has to be accurate enough to avoid unnecessary supply switches. A comparator with hysteresis is mostly used to gain noise-immunity. However if we take this hysteresis into account in formula 2.32, a hysteresis of 2.5% will already cause a power efficiency drop of 5%.

- The extra supply voltages need to be generated as well. So the efficiency of the DC-DC converter has to be added to the efficiency formula (2.32). State-of-the-art DC-DC converters tend to reach an 80% efficiency [May et al., 2001]. If the high voltage is generated a 3% efficiency drop can be calculated. For the generation of the low voltage, this decay will become 6%.

- The used class AB power amplifier needs to have a very high Power Supply Rejection Ratio (PSRR) since its supply rail is hard switched. The switching noise has to be suppressed towards the output to still achieve high linearity.

Due to these constraints it is believed that adding more power supplies to a DMT modulated signal class G power amplifier will not be beneficial from a power efficiency point of view.

The class H amplifier takes the adaptive supply rail approach further. In stead of hard switching, the supply voltage is progressively following the output signal. This increases the efficiency and relaxes the PSRR constraints for the basic class AB line driver. The supply voltage of the class AB line driver in the class H configuration is modulated directly by the input voltage. It can also be regarded as a class G power amplifier with an infinite amount of supply voltage sources.

The most common implementation is that of a class AB line driver with a fixed supply that is capacitive boosted with the input supply voltage [Bicakci et al., 2003]. Since the class H amplifier can work from one supply, no extra DC-DC converter is necessary. The necessary capacitors however need to be of high quality and are external components. The bill-of-materials will thus almost be equal. This technique is heavily used in high-performance, high efficiency audio power amplifiers [van der Zee and van Tuijl, 1998], for xDSL line drivers however the class H does not yet outperform the mature class AB designs yet.

5.3   Class D

5.3.1   Basic Class D configuration

A typical class D consists of three major parts: a Pulse Width Modulation (PWM) or Pulse Density Modulation (PDM) which transforms the analogue input signal into a switching signal, large output switches to deliver the current to the load and an output filter to get rid of the high frequent switching signals. A principle schematic is shown in figure 2.28. The distinction between synchronised and self-oscillating class D amplifiers is made whether the modulator is clocked or not. At first, only the synchronised type of class D power amplifiers is considered.

The major advantage of the class D principle is the use of a switching output stage. If the on resistance of the output switches are neglected, there is no simultaneous current through the active element and voltage standing over the active element. So no heat is dissipated in the switches, opening the opportunity for 100% efficiency. For its high efficiency the class D power amplifier is very important in the low power consumption audio application field like hearing aids, etc.. To use one as a line driver several extra considerations need to be taken into account :

- For audio signals the output filter does not need to be very steep for audio signals have a low bandwidth (<20 kHz) compared with the switching frequency (usually between 200 kHz and 500 kHz). Moreover the loud-

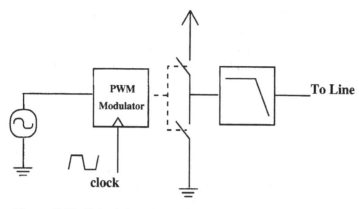

Figure 2.28: Principle schematic of a class D power amplifier

speaker set (and eventually also the human ear) inherently is a low pass filter for those switching frequencies.

- xDSL line drivers have relatively high bandwidths and very stringent out of band specifications. Therefor the output filter needs to be very steep or the over-switching ratio (OSR = mean switching frequency / signal bandwidth) has to be set high enough. The output filter needs to be a passive filter, so for line driver applications it will require many extra passive components, lowering the integrate-ability and increasing the cost of the complete driver.

- Another consequence of a higher switching frequency is that the switching losses will become more important in the overall power consumption. So much care needs to be taken to optimally choose the switching frequency in order to still gain efficiency.

From this brief analysis it is to be concluded that the use of a technology which already matured in the audio field as a line driver opens up a whole set of new design consideration.

5.3.2 Output Stage Considerations

On-resistance. The design of the output stage is a far from trivial task. From a power point-of-view several contributions need to weighed against each other. The first important parameter is the conduction loss due to the non-zero on resistance of a real switch. When integrating these switches, this will lead to bigger transistors. An alternative would be the use of a higher supply voltage. The load resistance can be transformed to higher values for the same output power. This problem becomes more pronounced when integrating a line driver in a low voltage technology.

Non-zero switching time. Another important efficiency degrading effect is the relative importance of the switching losses in the overall power budget. During a switching event, there is a current flowing through the switch, while a voltage over the switch is building up. This leads to extra power dissipation. For this the switching time needs to be as low as possible. This can be obtained by reducing the output transistor sizes, which is the opposite action from the previous degradation effect. Another approach would be reducing the mean switching frequency, in order to reduce the relative portion of the switching losses. For the class D to be used as a line driver, the large bandwidth could be problem. If the on-resistance of the switch is small, an almost short circuit is made between supply and ground during the transition, giving an extra power consumption. This shoot-through current can be cut by inserting a dead time between the switching events of both switches. This however leads to an important in-band cross-over distortion, since the output voltage is uncontrolled during the dead zone.

Dynamic power consumption. Closely connected with the previous observations is the dynamic power consumption. Huge switches tend to have large input capacitances which need to be charged and discharged at every switching event. Furthermore, in order to reduce the switching time, these large input capacitances need to be driven by buffer amplifiers whose power consumption will increase with decreasing output switching time and larger input capacitance of the final stage.

In-band signal integrity. The class D configuration of figure 2.28, shows an open loop class D power amplifier. Due to the switching nature of the output signal, feedback is not that trivial. If no feedback is foreseen, the in-band signal integrity is dependent on the power line integrity. Every spur on the supply rail will be directly transferred towards the output. Since power rail bounce is directly related by the power consumption, a regulator with enough decoupling should be added on the supply rail which has an equally strict linearity as the required output linearity.

Another source of distortion in an open loop class D line driver is the cross-over distortion which occurs during an output transition. The cross-over distortion does not only originate from a no shoot-through current circuitry, but also from a the inductance of the output filter. Since the current through an inductor can not stop immediately, the remanent inductive load current flows through a free-running diode. The diode voltage adds to the supply, so on the output a voltage occurs which is smaller than ground or bigger than the upper supply rail. These effects become more pronounced for higher switching frequencies, for their relative importance on the integrated output voltage (through the output filter) will increase.

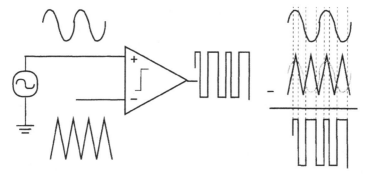

Figure 2.29: Principle schematic of a natural sampling, open loop modulator block.

Other Remarks. Other important specification for the design of a class D output stage are of more process technological nature. The diodes need to have fast recovery times and the technology should have a high latch-up resistance when these diodes conduct the remanent inductive current.

As it has already been pointed out, the use of higher output voltages is always beneficial for the design of a class D line driver. This can be accomplished by using a more expensive DMOS technology, which reduces the integrate-ability of the line driver with other system building blocks. Another approach could be the use of drain-source engineering in standard CMOS technologies [Sowlati and Leenaerts, 2002]. When a high voltage is not feasible, high currents will occur in the output path. Careful designs is mandatory in order to avoid 'hot spots' and electro-migration of the metal. Techniques from clock distribution networks need to be used in the layout of a segmented output switch to guarantee an equal current spread.

5.3.3 Modulation Schemes

Figure 2.29 shows the basic and most common open loop modulation technique, which is called natural sampling. It has been used for its simplicity and for the absence of any extra digital processing. Normally, no harmonic distortion is generated by this technique. The modulation however generates spurs at multiple harmonic offsets from the central clocking frequency. This limits the bandwidth of the system.

Another big disadvantage is the absence of a feedback loop, so all circuit non-idealities like output stage inaccuracies, non-linearity, timing errors or supply voltage ripple cannot be compensated. There exist techniques to allow the insertion of feedback to the natural sampling technique, but :

- The output needs to be filtered before it is fed back. The slew rate of this filtered output needs to be much smaller than the slew rate of the triangular wave. Otherwise extra higher harmonics will be generated driving the amplifier in a high frequent self-oscillating mode. The mean switching frequency is pushed to higher frequencies with all associated problems for the output stage.

- The consequence of this low corner frequency of the feedback filter is a low loop gain and thus a low compensation of output errors. This limits the performance improvement.

- The output filter cannot be made sharp enough to filter out the switching component completely, since the parasitic resistance limits the filter order. This remaining switching component will be amplified by the pre-amplifier and will arrive at the comparator were it most certainly will be out-phased with the original triangular wave. The intermixing products will generate in-band distortion.

- High frequent ripple picked up from the environment at the inputs of the comparator can cause erroneous switching of the class D amplifier which cannot be corrected by feedback, since it is filtered out in the correcting loop. The comparator and pre-amplifier therefor have to be shielded very carefully.

- Pulse amplitude errors are due to changes in the output load. These changes are very common in wire-line communications.

Another approach would be the use of digital modulation techniques like $\Delta\Sigma$-modulation to create a PWM signal [Dallago, 1997]. In this technique the errors are shaped towards higher frequencies. An external filter needs to filter out these high frequency errors. The major advantages of these techniques are

- The accuracy of the analogue components is largely relaxed, compared with the PWM modulator.

- Since the high frequency errors are randomised, the $\Delta\Sigma$ modulation generates less in-band substrate noise and thus less self-interference.

- Baseband feedback is inherent in the system, so in-band distortion is lower for the same clocking frequencies.

- For low signal bandwidths this can be combined with an Analog to Digital Converter (ADC) if the base signal is already digital [Philips et al., 1999].

Formula 2.33 calculates the Dynamic Range (DR) of a $\Delta\Sigma$ based switching power amplifier.

$$DR = \frac{3\pi}{2}(2n+1)\left(\frac{OSR}{\pi}\right)^{2n-1} \tag{2.33}$$

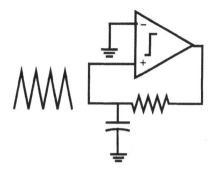

Figure 2.30: Basic relaxation oscillator schematic

The $\Delta\Sigma$-loop has order n and an over-switching ratio (OSR). The order should be kept as low as possible, since the order of the output filter needs to be at least one order higher than the $\Delta\Sigma$ order. So in order to meet the necessary linearity specifications, a higher OSR has to be chosen. This however, requires very high frequent high-power devices, which are not available in present semiconductor technologies.

5.3.4 Self-Oscillating Class D

The construction of a self-oscillating power amplifier (SOPA) originates from the observations of the natural sampling scheme of figure 2.29. The easiest way to generate a triangular wave is by designing a relaxation oscillator. The schematic of a basic relaxation oscillator is depicted in figure 2.30.

Since the building blocks of a relaxation oscillator are the same as the ones for the natural sampling modulation technique, a merge of the two circuits seems a logical step. The result of this is shown in figure 2.31. In the same figure the typical waveforms are plotted. They can be interpreted as follows: since the filtered version of a square wave is a triangular wave, a triangular wave is most likely to be found at the negative input of the comparator. Due to the feedback, the input signal can be considered as a shift in bias point, so a triangular approximation of the input signal resides at the negative input node. Since for in-band signals the system is in unity feedback and the loop filter is assumed to be linear, the in-band frequency content of the output square wave will be the same as that of the triangular wave and thus as the input-signal.

Although the circuit was firstly conceived to save components in the audio field, it became more popular due to its high-performance modulation. However, it's applicability for xDSL is far from trivial due to the high bandwidth and linearity constraints. This work is dedicated to this type of line drivers. The basic architecture will be further analysed and further improvements will be presented throughout this book. It will be shown that further improvements can be made by fully exploiting the properties of these hard non-linear systems.

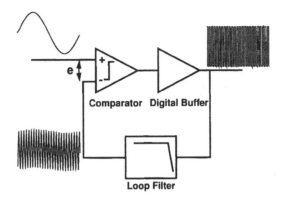

Figure 2.31: Block schematic of a self-oscillating power amplifier and its typical waveforms

5.4   Class K and other combined structures

The basic schematic of a class K power amplifier is depicted in figure 2.32. It consists of a class AB power amplifier whose output current is monitored. When the class AB's output current and thus also its power consumption exceeds a certain threshold, an extra current source is switched in. This switched driver thus delivers the main power, while the class AB is put in a feedback configuration to guarantee the required linearity. Much care has to be taken that the feed-forward connection towards the switched driver in combination with the feedback through the class AB power amplifier does not trigger a limit cycle oscillation. An unstable class K system will loose its linear properties. This type of line driver has been successfully proposed for audio applications [van der Zee and van Tuijl, 1998] , but its feasibility for xDSL needs to be proven yet. The biggest drawback will be the stability constraint in combination with the high bandwidth of an xDSL signal. The timing delay difference between the class AB signal path and the path through the current sensing circuitry and the switched driver will be a major bottleneck. The switching will also introduce spurious peaks at higher frequencies who need to be filtered out to meet the out-of-band specifications.

Another form of class K operation is presented in [Sæther et al., 1996]. In this design the output current is monitored, but instead of switching extra current, an extra class AB power amplifier is turned on. This technique enables rail-to-rail output swing with low distortion and a decrease of offset errors.

Other circuit topologies that make combinations of different type line drivers are also envisioned today [Sevenhans et al., 2002]. These 'Beauty and the Beast' topologies use two line drivers in parallel which are both driven by the same input signal. The 'Beauty' is a very linear class AB line driver that is put in a feedback configuration. The 'Beauty' needs to linearise the signal that is

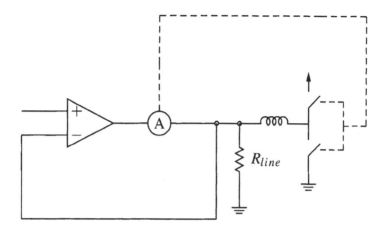

Figure 2.32: Basic circuit schematic of a class K power amplifier

put on the line by a more efficient 'Beast' line driver who will deliver the major part of the power budget. The major advantage of this configuration would be that it is easier to combine this structure with active back termination [Moons, 2003] than in a class K configuration. However no implementation of this concept has been presented up to now and it is believed that although the structure looks promising it will not lead to a more efficient line driver than the present state of the art class G power amplifier. This is due to :

- If the 'Beast' amplifier is a class B amplifier, meaning a class AB line driver with a too low or no quiescent current, the major part of the current will have to be delivered by the 'Beauty' since the voltage spikes occur rarely.

- If the 'Beast' is a highly linear class AB power amplifier with reduced headroom, i.e. it has a high efficiency for the rms signal but generates a lot iof clipping noise. The system is then comparable with a class G amplifier from its behaviour point-of-view but the power amplifier is double instantiated.

- The only 'Beast' left to give a higher efficient power amplifier is a switching class D amplifier. Much care has to be taken that the switching signal does not overdrive the inputs of the class AB line driver. The same stability constraints as for a class K power amplifier will limit its bandwidth. Power supply ringing and high frequent spurious suppression will be impossible since the switching signal needs to be filtered in the feedback of the class AB amplifier. The concept will have the same drawbacks as switching power amplifier in feedback configuration, see section 2.5.3.

For all these reasons, the 'Beauty and the Beast' topology will lead to the combination of a 'Beast' which is too beautiful and a 'Beauty' that fades away through hard labour.

6.   Conclusions

For more than a century, telecom operators have invested a huge amount of money in putting twisted pair wires in the ground. When the demand for broadband access to the Internet grew, new modulation techniques were necessary to reuse the unshielded telephone wires for these high bit rate applications. By making use of DMT-modulation the lossy channel could be utilised almost up to Shannon limit. It however became clear that this clever modulation was devised by system engineers who apparently didn't have the considerations for building an analogue front end in mind. Since DMT-signals have very high crest factors, they add a very stringent specification to the design next to all limitations connected with the bad channel.

This problem becomes the most pronounced in the design of a line driver for xDSL. Due to the high crest factor, the power amplifier is the major contributer to the overall power budget and is currently the limiting factor for the integration of more than 24 lines on a central office line termination board. Due to the heavy linearity specifications in combination with the high bandwidth, the class AB line driver is up to now the most used line driver in the field. By gradual improvements the class AB design has been pushed to its theoretical limit of 12% efficiency, which is actually unacceptable low.

New circuit techniques have been tried, like class G and H, and although they can reach a theoretical 50% efficiency limit, the techniques are still at the class AB power consumption level. Other techniques like class K and combined line drivers have been proposed but none of them have been realized so far.

If one really goes for a highly efficient power drive, one needs to consider a switching type line driver. Traditional, synchronous class D power amplifiers will run into the low distortion constraint, certainly for DSL types that go to very high signal frequencies. A Self Oscillating Power Amplifier (SOPA) has been briefly presented. Due to its non-linear continuous time nature it possesses peculiar properties that enable the construction of a highly linear, high efficiency line driver that is able to cope with the stringent xDSL specifications. In the following of this book, this SOPA structure will be further analysed and realised to prove its use.

Chapter 3

DESCRIBING FUNCTION ANALYSIS

IN this chapter the mathematical foundations for the describing function method are briefly explained. Since the SOPA is a hard non-linear system, traditional control theory is not applicable for the design and analysis of the system. The major differences between linear and non-linear control theory are presented in a first section. Before the actual analysis can get started the available tool set needs to be presented. Since traditional linear analysis techniques are not valid anymore in a non-linear system.

In linear control theory, the frequency response method, also called Fourier analysis, is the most important analysis method. This is not only due to the various graphical methods that have been developed to analyse the linear control system like Bode, Nyquist or root locus plots but also due to the fact that the frequency response methods are easily extended when going to higher order methods. Also the methods provide physical insight, because there is a direct connection with the physical configuration. This is due to the fact that a sinusoidal wave is a base-function for every linear system.

The describing function method provides a quasi-linearisation of the non-linearity in the frequency domain, so traditional linear control theory can be used for the design. In this chapter, the basics of this theory will be further explained. The method will then be the basic tool set used in the complete analysis of a SOPA system in the following chapter.

1. Non-linear Systems

1.1 A Signal Point-of-view

Figure 3.1 shows a possible classification on the nature of the signals occurring at the output of a system. The most natural way to present signals is in a pure analogue or a pure digital fashion. Digital signals introduce truncation

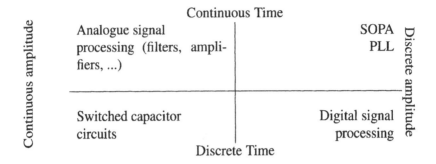

Figure 3.1: Classification of a system based on the signal properties

errors in both the amplitude as in the time domain. These errors are regarded as a noise source, i.e. iquantisation noise, in a communication system.

Two intermediate regions are also in use in present electronic systems as indicated in figure 3.1. Since the signals in those type of configuration have one continuous dimension, it is possible to construct circuits that don't have quantisation noise in the signal bandwidth. The SOPA is located in the discrete amplitude, continuous time corner since its output signal is switching but not synchronised.

1.2   Linear versus Non-Linear

Dynamical systems can be described by their state-space model :

$$x(t) \;=\; f(t, t_0, x_0, u_{(t_0,t)}) \tag{3.1}$$
$$y(t) \;=\; g(t, x, u(t)) \tag{3.2}$$

Equation 3.1 denotes the property that the state variables of a system depend on their initial value x_0 at time t_0 and the input signal $u(t)$ in the half closed interval $(t_0, t$. This equation is called the state equation. The system outputs $y(t)$ are the result of the output equation (3.2) and are only dependent on the state variables x and the input at time t. The system is called autonomous if (3.1) has no input. The input-output relation ρ can be directly calculated by filling in (3.1) in (3.2) :

$$y(t) = g(t, f(t, t_0, x_0, u_{(t_0,t)}), u(t)) = \rho(t, t_0, x_0, u_{(t_0,t)}) \tag{3.3}$$

Linear systems fulfil the following superposition principles :

1 The decomposition property :

$$\rho(t, t_0, x_0, u_{(t_0,t)}) = \rho(t, t_0, 0, u_{(t_0,t)}) + \rho(t, t_0, x_0, 0) \tag{3.4}$$

2 linearity of the zero-state response :

$\forall a, b \in \mathcal{R}, \forall u_1, u_2 \in$ input signals defined over $(t_0, t$

$$\rho(t, t_0, 0, au_1 + bu_2) = a\rho(t, t_0, 0, u_1) + b\rho(t, t_0, 0, u_2) \quad (3.5)$$

3 linearity of the zero-input response :

$\forall a, b \in \mathcal{R}, \forall x_{01}, x_{02} \in$ initial conditions

$$\rho(t, t_0, ax_{01} + bx_{02}, 0) = a\rho(t, t_0, x_{01}, 0) + b\rho(t, t_0, x_{02}, 0) \quad (3.6)$$

A non-linear system is a system that does not obey one or more of the above linearity properties. Therefor the big advantage of linear systems does no longer apply for non-linear systems, namely the fact that a linear system can be split up in manageable parts and the results of the analysis of several building blocks can be linearly combined to form the overall solution.

A parametric curve over time in the state-space is called a trajectory. By the uniqueness of solution of an autonomous dynamical system, there is one and only one trajectory through every point in state-space. As a consequence of this, trajectories cannot intersect. The graphical representation of these trajectories is called the phase-portrait of a system. An important property for non-linear systems is the existence of a wider range of possible phase-portraits. For example a non-linear system can exhibit limit cycles. These are closed isolated trajectories (orbits). With isolated orbits, trajectories are meant to which trajectories in the neighbourhood converge or diverge. A linear system like an ideal undamped pendulum does have closed orbits but this will lead to a phase-portrait with only concentric trajectories. If a small perturbation is applied, the oscillation will continue on another orbit. The special properties of limit cycling will be further explained in the next chapter where the SOPA is further elaborated.

1.3   Hard- versus Soft-Non-Linearity

The nature of the non-linearity will determine the analysis method which will provide the most correct / best interpretable results. This classification, however cannot be made without regarding the amplitude of the signals in the system[1]. With a soft non-linearity, a linearity is meant that can be sufficiently linearised for the applied signals. If the error made by a Taylor series expansion of the non-linearity is small enough for the signals at its inputs, the Volterra series approach can be utilised to analyse the circuit [Wambacq and Sansen, 1998].

---

[1]There are textbooks who discriminate hard/soft non-linearities by demanding continuity. We will adopt a less stringent criterion.

The modulation of the SOPA amplifier is performed by a comparator. If this comparator has an infinite gain, it is by default a hard non-linear element in a continuous time loop. Even if, and in reality this will be, the comparator has finite gain, the loop is designed so that it will lead the comparator into saturation. A saturation event cannot be approximated by a limited number of terms of a Taylor series, so it is still considered to be a hard non-linearity.

1.4   Solution Methods

Since there are no direct, general methods to solve a non-linear system, several techniques are available. The choice of technique depends on the nature of the system and on which characteristic one wants to derive. The following non-exhaustive list, points out some other methods worth to consider

Prototype testing : since the outcome of any analysis is as good as the model used, prototyping is necessary to have a formal verification of the design methods. Prototyping, however is very costly and time consuming. The prototypes used in this work will be presented in chapter 6

Numerical simulations : due the continuing increase in computational power, numerical simulations become faster and more usable. The big drawback however is that they do not really contribute to provide insight in the systems behaviour. Chapter 5 will discuss these methods into more details.

Closed form solutions : only for a very few differential equations, a closed form solution can be found. If it is applicable for the system under design, this is the favourable method.

Lyapunov's direct method : is one of the most important methods to determine the systems stability[2]. The method is based on finding a Lyapunov function $V(x)$ on a system's trajectory, which is comparable with a potential energy function. The trajectories into the vicinity of a fixed point, should have a Lyapunov function whose first derivative is negative, for the fixed point to be stable.

Tsypkin's method : for systems whose only non-linearity is a ideal comparator, Tsypkin has constructed an exact method to calculate the systems performance [Cypkin et al., 1962]. The method can be used to calculate synchronisation, limit cycle characteristics, sub-harmonics and the response on a forced signal. The method gives exact results which are easily interpretable by the graphical methods. The limitation to only ideal compara-

---

[2]For non-linear systems, oscillation does not imply instability, due to the existence of limit cycles. An oscillation that returns to the same frequency/amplitude after a small perturbation is a stable oscillation mode.

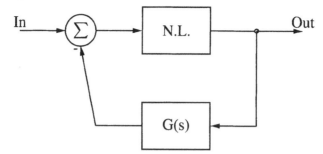

Figure 3.2: Basic system configuration that can be used with the describing function analysis.

tors, the complexity of the calculation and the incompatibility with linear analysis methods damp its popularity.

2. The Describing Function Method
2.1 The Goal

The goal of the describing function method is to provide the designer with a mathematical tool that allows the use of frequency domain methods in systems that contain hard non-linear blocks. Note that the method will only work for systems which have localised non-linearities. so the non-linearity can be grouped in a separate block while the other building blocks have linear representations. Also the non-linearity needs to be odd and time-invariant.

In general these systems could be simplified to the form of figure 3.2. The describing function will replace the non-linearity (N.L.) by a quasi-linear block so that the input-output relation of the systems is approximately equal to the original system behaviour. The term quasi-linear is used, since a describing function will be dependent on the applied input signal. A direct consequence of this, is that there does not exist one describing function for a certain non-linearity. A describing function is always determined by both the non-linearity and the signals in the system.

2.2 Unified Theory of the Describing Function Method

The describing function can be graphically presented as in figure 3.3. The input signal is split in several input signals $x_i(t)$. These inputs have their own weighting function in the filtered output (= their own describing function). After summation the obtained approximation $y_a(t)$ needs to match the output $y(t)$ as close as possible.

As close as possible, has been implemented in the describing function method as a minimisation of the mean-squared error :

$$\overline{e(t)^2} = \overline{y_a(t)^2} - 2\overline{y_a(t)y(t)} + \overline{y(t)^2} \tag{3.7}$$

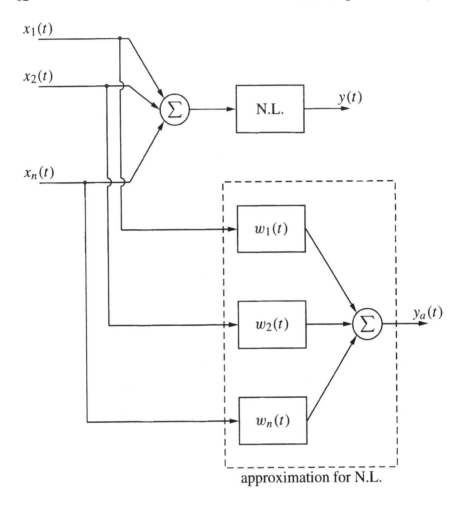

Figure 3.3: General linear approximation of a non-linearity

The weighting functions $w_i(t)$ need thus be chosen, so that the mean-squared error is minimal. If the various input components are statistically independent, the correlation between the different input signals is zero and drops from the least square error calculation. This will mean that the input signals $x_i(t)$ will have different characteristics, so only one bias signal can be part of the set of input signals $x_i(t)$. If more signals with an equal nature are present, they should influence the calculation of each other weighting function. The resulting describing functions will thus be dependent on each input signal, but be only a gain block for the input signal it is related to. If the input signals are uncorrelated, the weighting functions fulfil the following condition [Gelb and

Vander Velde, 1968]:

$$\int_0^\infty w_i(\tau_2)\overline{x_i(t)x_i(t+\tau_1-\tau_2)}d\tau_2 = \overline{y(t)x_i(t-\tau_1)} \qquad \tau_1 \geq 0, i = 1,2\ldots,n$$

$$(3.8)$$

The weighting functions can thus be calculated by equating the cross correlation between the filtered input and the input itself with the cross correlation between the output of the non-linearity and the specified input. For the most common signals such as : bias, sinusoidal and Gaussian signals this criterion can be further simplified and easy representations for the specific describing functions have been obtained.

2.3 Limitations of the Method

Since the signal at the input has to be guessed beforehand, it limits the possible system architectures when feedback structures are in use. Referring to figure 3.2, if the describing function used is a sinusoidal input describing function (see section 3.2.4.1), the input should be almost sinusoidal. The linear $G(s)$ block should thus have a low-pass characteristic to suppress the higher harmonics when the output is fed back. This is called 'the filter criterion' and is the most important criterion for validity of the results.

The describing function method is an approximation of the real characteristic. Its validity cannot be guaranteed beforehand, although there exist methods that predict the error [Bergen et al., 1982, Blackmore, 1981] during analysis. Both provide a mathematical interpretation for the filter criterion.

2.4 The Sinusoidal Input Describing Function

2.4.1 Calculation

This type of describing function is the most commonly used one, so it is sometimes referred to as the Describing Function (DF). For a sinusoidal signal $x_i(t) = A\sin(\omega t + \theta)$ with A and ω determined and the phase θ be a uniformly distributed random variable between 0 and θ. The left hand side of (3.8) can than be calculated by first calculating the cross-correlation of a sinusoidal input :

$$\overline{x_i(t)x_i(t+\tau)} = \frac{A^2}{2}\cos(\omega\tau) \qquad (3.9)$$

Which will lead to :

$$\int_0^\infty w_A(\tau_2)\overline{x_i(t)x_i(t+\tau_1-\tau_2)}d\tau_2 =$$

$$\frac{A^2}{2}\cos(\omega\tau_1)\int_0^\infty w_A(\tau_2)\cos(\omega\tau_2)d\tau_2 + \frac{A^2}{2}\sin(\omega\tau_1)\int_0^\infty w_A(\tau_2)\sin(\omega\tau_2)d\tau_2$$

(3.10)

The right hand side of (3.8) will be :

$$\overline{y(t)x_i(t-\tau_1)} = A\cos(\omega\tau_1)\overline{y(0)\sin(\theta)} - A\sin(\omega\tau_1)\overline{y(0)\cos(\theta)}$$ (3.11)

Satisfaction of (3.8) for all non-negative values of τ_1 requires :

$$\frac{A}{2}\int_0^\infty w_A(\tau_2)\cos(\omega\tau_2)d\tau_2 \;=\; \overline{y(0)\sin(\theta)}$$ (3.12)

$$\frac{A}{2}\int_0^\infty w_A(\tau_2)\sin(\omega\tau_2)d\tau_2 \;=\; -\overline{y(0)\cos(\theta)}$$ (3.13)

The describing function that can be derived from solving this system of equations will be the following complex gain :

$$N_A \;=\; n_p + jn_q$$ (3.14)

$$n_p \;=\; \frac{2}{A}\overline{y(0)\sin(\theta)}$$ (3.15)

$$n_q \;=\; \frac{2}{A}\overline{y(0)\cos(\theta)}$$ (3.16)

For non-linearities which are static and single valued, so that $y(0)$ us given unambiguously in terms of $x(0)$, this quadrature gain is always zero. For non-linearities which are static, meaning they only depend on the actual input voltage and not on the derivatives of the input, the cross-correlations can be evaluated by :

$$N(A) = \frac{j}{\pi A}\int_0^{2\pi} y(A\sin(\theta))e^{-j\theta}d\theta$$ (3.17)

One can clearly observe the resemblance with a Fourier transform. This can be directly coupled with its physical meaning. The describing function $N(A)$ is the gain a sinus has to be multiplied with to obtain the output rms amplitude at the input frequency. The DF of a memoryless[3] non-linearity clarifies this even more, if one considers the first harmonic term of a Fourier series expansion.

$$N(A) = \frac{2}{\pi A} \int\limits_{-\pi/2}^{\pi/2} y(A \sin(\theta)) \sin(\theta) d\theta \qquad (3.18)$$

2.4.2 Use of the single sinusoid DF

single sinusoid DF The DF is thus applicable in systems where only one sinusoidal signal is applied to the system. Since the cross correlation of two sine waves does not go to zero, the approximations in the calculation of the single sinusoidal describing function does no longer hold. So, if more sine waves occur in the signal path and they are not sufficiently suppressed, the single sinusoidal describing function will lead to erroneous results. For this type of systems the Two Sinusoid Describing Function (TSIDF) has to be used as will be explained in section 3.2.5.

Typical application of the DF is in the determination of limit cycles in autonomous systems. Since no input is applied to such a system, the only possible signal, next to chaotic oscillation which lies beyond the scope of this method, will be a limit cycle oscillation. This kind of behaviour is easily calculated by the Barkhausen criterion. For the standard system of figure 3.2 this will lead to solving the following complex equation :

$$G(j\omega)N(A) + 1 = 0 \qquad (3.19)$$

Note that this equation can have more than one solution, since it is a system of complex equations of higher order. The DF method does not always provide a single value solution. If more solutions are obtained with positive amplitude and frequency, the stability of each solution needs to be checked. A sufficient condition for the limit cycle stability can be calculated by formulating the Barkhausen criterion (3.19) explicitly as a complex equation :

$$U(A, \omega) + jV(A, \omega) = 0 \qquad (3.20)$$

A solution (A_i, ω_i) will be the amplitude and pulsation of a stable limit cycle if :

$$\frac{\partial U}{\partial A} \frac{\partial V}{\partial \omega}\bigg|_{A_i,\omega_i} - \frac{\partial U}{\partial \omega} \frac{\partial V}{\partial A}\bigg|_{A_i,\omega_i} > 0 \qquad (3.21)$$

---

[3]i.e. a single-valued non-linear characteristic ; multi-valued characteristics are said to possess memory (Cf. hysteresis).

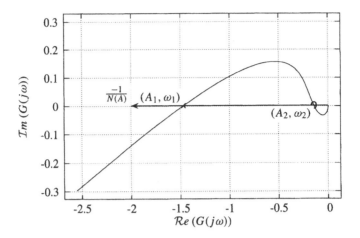

Figure 3.4: Illustration of the graphical method to determine stability

The derivation of this stability condition in a polar form is subject of appendix A. This criterion is sometimes referred to as Loeb's criterion [Atherton, 1975]. The use of the describing function method will lead to a very accurate prediction of the oscillation mode as long as $G(s)$ fulfils the filter criterion.

Another application will be forced systems that do not exhibit limit cycle oscillation but contain a non-linear element, like e.g. an analogue control loop in which an element is pushed into saturation.

2.4.3   The Modified Nyquist Plot

The Barkhausen criterion of (3.19) can be visualised by plotting the curves of the linear part of the system $G(s)$ in the complex plane together with the function $-1/N(A)$. The intersection between both curves determines the solutions of the Barkhausen criterion. This is illustrated in the example of figure 3.4. The solution at the origin is impossible since it represents a limit cycle with infinite pulsation and zero amplitude. The method however still proposes two solutions.

The stability can be determined by regarding figure 3.4 as a regular Nyquist plot with modified x-axis. If $G(s)$ does not have poles in the right half-plane, the (-1,0) should not be encircled for the closed loop system to be stable. At the limit cycle solutions (A_1, ω_1) and (A_2, ω_2) the linear Nyquist model would be a polar plot of the loop gain through the (-1,0) point. The corresponding closed loop solution is a sustained oscillation. If the system is driven out of his equilibrium point by a small increase or decrease in amplitude $\pm\delta A$, the limit cycle is said to be stable if the closed loop system evolves to an oscillatory state with same frequency and amplitude. For the solution around (A_1, ω_1), an increase in amplitude will drive the -1 point of the linearised Nyquist plot to the left of the limit cycle solution. The closed loop system does no longer encircle the (-1,0) point. The closed loop system is thus stable and the oscillation will be suppressed. The amplitude will thus decrease and the system evolves towards it original state. If the amplitude of the limit cycle oscillation around (A_1, ω_1) is slightly decreased, the (-1,0) point shifts towards the origin. It is now encircled by the $G(s)$ polar curve. Following linear theory, the system becomes unstable and the amplitude will grow. It can be concluded that the (A_1, ω_1) limit cycle is stable. An analogue reasoning will result in the instability of the other limit cycle oscillation.

A direct interpretation is by ordering the intersection points from $-\infty$ towards the origin. Every odd numbered solution will provide a stable limit cycle oscillation.

2.4.4   Important Describing Functions

As an example, the describing function of the saturation function will be calculated. The saturation function is a linear gain block with gain factor m. From a certain input value δ on, the output saturates to the value $m\delta$. For the calculation of the describing function, two cases need to be addressed first. If the amplitude of the signal under consideration A is smaller than δ, no non-linearity is observed. The describing function equals its linear gain m. For the calculation of the case where $A > \delta$, formula 3.18 needs to be used since the saturation function is static and memoryless. So until a certain phase $-\theta_1$ the value of the input function is below $-\delta$ and saturation occurs. The same occurs for a phase value θ_1 for which the signal exceeds δ. Since both the signal and the non-linearity are symmetric around the origin, the absolute value of the crossing phases can be taken equal. Equation 3.18 can then be transformed into :

$$N(A) = \frac{2}{\pi A} \left[\int_{-\frac{\pi}{2}}^{-\theta_1} -m\delta \sin(\theta)d\theta + \int_{-\theta_1}^{\theta_1} mA \sin^2(\theta)d\theta + \int_{\theta_1}^{\frac{\pi}{2}} m\delta \sin(\theta)d\theta \right] \tag{3.22}$$

The integrals can be solved to :

$$N(A) = \frac{2}{\pi A} \left[-2m\delta \cos(\theta_1) + mA\theta_1 + \frac{mA \sin(2\theta_1)}{2} \right] \tag{3.23}$$

Since by definition θ_1 is the solution of the following equation :

$$A \sin(\theta_1) = \delta , \tag{3.24}$$

the describing function will ultimately become :

$$N(A) = \frac{2m}{\pi} \left[\arcsin\left(\frac{\delta}{A}\right) + \left(\frac{\delta}{A}\right) \sqrt{1 - \left(\frac{\delta}{A}\right)^2} \right] \qquad (3.25)$$

An ideal comparator can be regarded as a saturation function with infinite gain. Its describing function can thus be easily derived from (3.25) by taking the limit $m \to \infty, \delta \to 0, 2m\delta \to V_{DD}$:

$$N(A) = \frac{2V_{DD}}{\pi A} \qquad (3.26)$$

A comparator which exhibits rectangular hysteresis is probably the most important non-linear function with memory. The describing function for a comparator with hysteresis from $-\delta$ to δ will thus be complex and is given by :

$$N(A) = \frac{2V_{DD}}{\pi A} \sqrt{1 - \left(\frac{\delta}{A}\right)^2} - j\frac{2V_{DD}\delta}{\pi A^2} \qquad (3.27)$$

2.5   The Two-Sinusoid-Input Describing Function

2.5.1   Calculation

From the general introduction of the describing function method, it can be understood that if two sinusoidal signals are applied to a system containing a nonlinearity, the describing function method will provide two describing functions which are dependent on both input signals. The gain of a sine wave with amplitude A in the presence of another input sine wave with amplitude B will be noted as $N_A(A, B)$. The general description of the TSIDF follows naturally from the calculations of the weighting coefficients :

$$N_A = \frac{1}{2\pi^2 A} \int_0^{2\pi} d\theta_A \int_0^{2\pi} d\theta_B \, y \, (A \sin(\theta_A) + B \sin(\theta_B)) \sin(\theta_A) \qquad (3.28)$$

$$N_B = \frac{1}{2\pi^2 B} \int_0^{2\pi} d\theta_B \int_0^{2\pi} d\theta_A \, y \, (A \sin(\theta_A) + B \sin(\theta_B)) \sin(\theta_B) \qquad (3.29)$$

To get rid of the double integral, a method based on the Bessel series expansion has been derived by Gibson and Sridhar [Gibson and Sridhar, 1963]. To use the method the non-linearity function has to rewritten as an inverse Fourier transform[4] :

$$y(x) = \frac{1}{2\pi} \int\limits_{-\infty}^{\infty} Y(ju)e^{jux} du \qquad (3.30)$$

If the x is replaced by the input signal $x = A\sin(\omega t) + B\sin(\gamma \omega t)$, the integral transformation can be rewritten as :

$$y(x) = \frac{1}{2\pi} \int\limits_{-\infty}^{\infty} Y(ju)e^{jA\sin(\omega t)}e^{jB\sin(\gamma \omega t)} du \qquad (3.31)$$

By replacing the $e^{jB\sin(\gamma \omega t)}$-terms by their Bessel series expansion and taking the first harmonic term, the first harmonic gain, being the TSIDF can be calculated as follows :

$$N_A(A, B) = \frac{j}{\pi A} \int\limits_{-\infty}^{\infty} Y(ju) J_0(Bu) J_1(Au) du \qquad (3.32)$$

$$N_B(A, B) = \frac{j}{\pi B} \int\limits_{-\infty}^{\infty} Y(ju) J_0(Au) J_1(Bu) du \qquad (3.33)$$

$$(3.34)$$

One can conclude that the frequency independent TSIDF for ech input component is functionally identical. That is

$$N_A(A, B) = N_B(B, A) \qquad (3.35)$$

2.5.2 Use of the TSIDF

Probably the most important use, certainly in this work, for the TSIDF is the calculation of the input response of a forced limit cycling system. The input signal is the second sine wave, since the limit cycle oscillation cannot be neglected.

Another important use of the TSIDF is the discrimination of multiple stable limit cycles. If a non-linear system has more than one stable limit cycle, it is important to know which one will occur in a physical system. For this, the TSIDF can be used to calculate the loop gain of the system for one limit cycle frequency while oscillating in the other stable mode and reversely. The mode with the highest loop gain will be dominant in the system, since a small perturbation on the other mode will lead to an unstable system.

---

[4]This is the general integral transform, not the special meaning as a transformation from the frequency domain to the time domain

Other uses include the occurrence of sub-harmonic oscillations, the study of a non-linear system's response on a two-tone test or transient responses of linear systems who are driven into saturation.

2.5.3   Some important TSIDF functions

As an example, the calculation of the TSIDF of an ideal comparator is given. The characteristic needs to be changed to

$$y(x) = \begin{cases} -De^{\sigma x} & x < 0 \\ 0 & x = 0 \\ De^{\sigma x} & x > 0 \end{cases} \tag{3.36}$$

to make it Fourier transformable. The resulting Fourier transform then becomes :

$$Y(ju) = \lim_{\sigma \to 0} \left(\int_{-\infty}^{0^-} -De^{\sigma x} e^{-jux} dx + \int_{0^-}^{0^+} 0 dx + \int_{0^+}^{\infty} De^{\sigma x} e^{-jux} dx \right) \tag{3.37}$$

$$= \lim_{\sigma \to 0} \left(\frac{-j2Du}{\sigma^2 + u^2} \right) \tag{3.38}$$

$$= \frac{2D}{ju} \tag{3.39}$$

The TSIDF can then be calculated as the result of the following integral by filling in (3.39) in (3.32):

$$N_B(A, B) = \frac{2D}{\pi B} \int_{-\infty}^{\infty} \frac{J_0(Au)J_1(Bu)}{u} du \tag{3.40}$$

The improper integral is from the Weber-Schafheitlin type [Luke, 1962] and can thus be calculated as follows:

$$\int_{0}^{\infty} t^\lambda J_\mu(at) J_\nu(bt) dt =$$

$$\begin{cases} \dfrac{(b/a)^\nu (a/2)^{\lambda-1} \Gamma\left(\frac{\mu+\nu-\lambda+1}{2}\right)}{2\Gamma(\nu+1)\Gamma\left(\frac{\mu-\nu+\lambda+1}{2}\right)} \, {}_2F_1\left(\frac{\mu+\nu-\lambda+1}{2}, \frac{\nu-\mu-\lambda+1}{2}; \nu+1; \left(\frac{b}{a}\right)^2\right) \\ \qquad\qquad \text{for } 0 < b < a \\[4pt] \dfrac{(a/b)^\mu (b/2)^{\lambda-1} \Gamma\left(\frac{\mu+\nu-\lambda+1}{2}\right)}{2\Gamma(\mu+1)\Gamma\left(\frac{\nu-\mu+\lambda+1}{2}\right)} \, {}_2F_1\left(\frac{\mu+\nu-\lambda+1}{2}, \frac{\mu-\nu-\lambda+1}{2}; \mu+1; \left(\frac{a}{b}\right)^2\right) \\ \qquad\qquad \text{for } 0 < a < b \end{cases} \tag{3.41}$$

Which results in the following describing function :

$$
N_B(A, B) = \begin{cases} \frac{2D}{\pi B}\left(\frac{B}{A}\right){}_2F_1\left(\frac{1}{2}, \frac{1}{2}; 2; \left(\frac{B}{A}\right)^2\right) & \text{for } 0 < B < A \\[2ex] \frac{4D}{\pi B}{}_2F_1\left(\frac{1}{2}, -\frac{1}{2}; 1; \left(\frac{A}{B}\right)^2\right) & \text{for } 0 < A < B \end{cases} \tag{3.42}
$$

By expanding the hypergeometric series, one finds:

$$
N_B(A, B) = \begin{cases} \frac{2D}{\pi B}\left(\frac{B}{A}\right)\left[1 + \frac{1}{8}\left(\frac{B}{A}\right)^2 + \frac{3}{64}\left(\frac{B}{A}\right)^4 + \frac{25}{1024}\left(\frac{B}{A}\right)^6 + O\left(\left(\frac{B}{A}\right)^8\right)\right] \\[1ex] \qquad\qquad \text{for } 0 < B < A \\[2ex] \frac{4D}{\pi B}\left[1 - \frac{1}{4}\left(\frac{A}{B}\right)^2 - \frac{3}{64}\left(\frac{A}{B}\right)^4 - \frac{5}{256}\left(\frac{A}{B}\right)^6 + O\left(\left(\frac{A}{B}\right)^8\right)\right] \\[1ex] \qquad\qquad \text{for } 0 < A < B \end{cases}
$$

$$\tag{3.43}$$

If $B \ll A$, (3.43) leads to the following describing function :

$$
N_B(A, B) \approx \frac{2D}{\pi A} = \frac{N(A)}{2} \tag{3.44}
$$

Using (3.35), the describing function for the sine wave with amplitude A can be derived from (3.43) for the same case where $B \ll A$.

$$
N_A(A, B) \approx \frac{4D}{\pi A} = N(A) \tag{3.45}
$$

The following conclusions can be drawn :

- The gain of the non-linear system for the small amplitude signal B will be independent on B itself. It can be stated that the large amplitude sine wave A linearises the non-linearity. Another point-of-view could be that the signal with amplitude A is the dither signal for the other sine wave.

- The describing function for the B sinusoidal input is half the single input describing function. If the A signal is a limit cycle oscillation, the $N(A)$ amplification factor is fixed by the Barkhausen criterion. The closed loop transfer function for the forced signal will thus be fixed by the limit cycle oscillation.

- The TSIDF for the bigger signal equals the single sinusoidal input describing function. So for the larger amplitude sine wave, the behaviour of the system equals the behaviour of a system without the smaller amplitude signal.

Other TSIDF functions worth mentioning include the comparator with finite gain m and saturation level $m\delta$:

$$N_B(A, B) = \frac{2m\delta}{\pi\delta B} \int\limits_{-\infty}^{\infty} \frac{\sin(\delta u)}{u^2} J_0(Au) J_1(Bu) du \qquad (3.46)$$

and the comparator with a dead zone with a width of δ:

$$N_B(A, B) = \frac{2D}{\pi B} \int\limits_{-\infty}^{\infty} \frac{\cos(\delta u)}{u} J_0(Au) J_1(Bu) du \qquad (3.47)$$

2.6   Other Describing Functions

Other important input signals related describing functions [Gelb and Vander Velde, 1968] include:

Three sinusoidal-input describing functions can be used to determine the (sub-)harmonics in a limit cycling system.

Dual-input Describing Function is the describing function related to a bias plus single sinusoidal input signal. dual input DF

Random Input Describing Functions can be used to calculate the transfer functions for several (white) noise sources in a non-linear system. It can also be used to describe random dither signals.random input DF

3.   Conclusions

Since a self-oscillating class D amplifier is a continuous time system with a hard non-linear building block (i.e. the comparator), the study of non-linear systems is mandatory to understand the working principles of the SOPA amplifier. Since the superposition theorem no longer holds, the system cannot be divided into parts to ease analysis. Furthermore, special phenomena like limit cycle oscillation can occur in non-linear systems.

In this chapter the basic tool set to handle non-linearities in a feedback system are briefly touched. The describing function splits up a signal in its basic components and a linear filter is constructed for every input signal. These quasi-linear filters (=describing functions) are constructed in such a way that the linear outputs meets the non-linear output in a least square error sense. The filters are quasi-linear, since they depend on the input signal itself.

The most important describing functions : the single sinusoid and two sinusoidal input describing function are further elaborated. An important feature of the TSIDF is that it is able to create a signal independent gain for a small sine wave in the presence of a larger sinusoidal signal. This gives a theoretical background for the dithering in a forced limit cycling system. This will be further elaborated for the SOPA in the next chapter.

Chapter 4

BEHAVIOURAL MODELLING OF
SELF OSCILLATING POWER AMPLIFIERS

IN this chapter a behavioural model for a special type of self-oscillating class D power amplifier, the SOPA is presented. We start by presenting a general reference model from which every possible SOPA implementation can be derived. The emphasis in this chapter is on the construction of mathematical models, as a preparation on the following chapters that will describe the practical implementation into more details. This chapter can thus not only be read as a modelling chapter that will aid a designer to optimise a SOPA-design, but also as an extensive example of non-linear system design. The methods developed in this chapter to model the frequency response and distortion of a hard non-linear system are also applicable to other architectures. Also the oscillation pulling, which is one of the most important properties of a SOPA is modelled in this chapter, and can also be used in the design of other systems in which the attraction of oscillators is (un)wanted.

The chapter is organised as follows : we start with a short introduction in which the reference model is explained. All references to a SOPA in the continuing of this book will denote a structure that is constructed in that way. Also a state space model is constructed that will be used in the numerical verifications of the obtained models. The remaining of the chapter is divided in two major parts. Firstly the complete analysis is performed for a zeroth order SOPA. For its lower complexity, the zeroth order SOPA is a good starting point to explain the major parameters and analysis techniques. The basic limitations of this structure will be explained and afterwards in a second part the solution presented by going to higher order amplifiers will follow the same analysis path.

In every part, the same structure is used to calculate the performance. The goal is to create non-iterative performance functions. In this way, for a given parameter set, the performance can be calculated in straightforward fashion.

For every calculation, the followed assumptions will be early explained. The modelling starts with the calculations of the limit cycle frequency and amplitude, since these will be performance determining parameters. Next, oscillator pulling is described and limits for various line conditions will be given. The forced system equations will lead to a model for the distortion due to the non-linear comparator characteristic and the amplifiers bandwidth.

Throughout this chapter much attention will be paid on gaining insight in the system by commenting the obtained results and considerations using various graphical methods.

1. Reference Model

1.1 General Description

In this section, the reference model of a SOPA is presented. The reference model depicts a generic structure from which every SOPA structure can be defined. The reference model is in its basic conception a pure mathematical model describing the architecture of a SOPA-amplifier. Although the primal goal in this chapter is to construct a mathematical model, the construction and visualisation of the model is implementation-driven, i.e. a basic implementation can be mapped directly to the model. Also, several physical limitations like limits of the line impedances, the basic properties of a signal transformer, etc. are taken into account from the conception of the model on. This enables us to generate manageable models and system equations. The obtained models however can be extended with various building block non-idealities to span the different corners of the design space. This will also be illustrated in this text.

The reference model of a general SOPA is depicted in figure 4.1. The reference model consists of two basic SOPA building blocks in a bridge configuration. The two SOPAs are connected using the line-transformer, giving galvanic isolation towards the line. Since this galvanic isolation is mandatory in wire-line communications to avoid ground loops, the inclusion of the line transformer in the SOPA system does not increase the number of costly external components.

The basic SOPA building block is depicted in figure 4.2. It consists of a feedback loop filter of order m with a cut-off frequency of f_{fil} and n integrators in the forward path. The unit-gain frequency of the integrators is f_{int}. A SOPA with n integrators will be called an nth order SOPA. Note that n can also be zero, while m is defined to be larger than 1. In case the SOPA is of zeroth order, the input is directly connected to the positive input of the comparator, while the output of the loop filter is connected via the gain β_1 to the comparator's negative input. From figure 4.2 every possible SOPA configuration can be derived by selecting the number of integrators n and the loop filters order n.

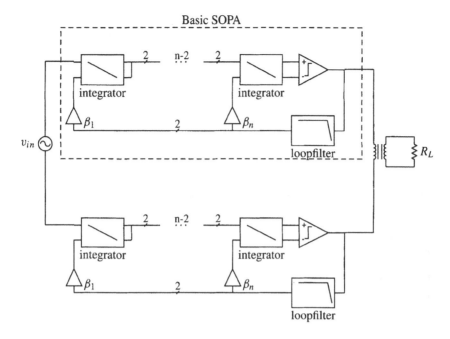

Figure 4.1: Reference model used in the behavioural calculations

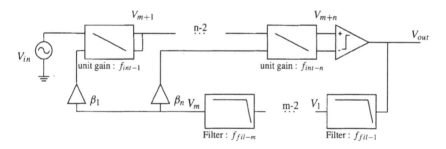

Figure 4.2: General basic self-oscillating power amplifier building block.

The output of the comparator is filtered and by various amplification factors β_i fed back to the various integrators.

1.2   State-Space Equations

While the frequency domain equations for the linear part can be derived straightforward from simple block schematic calculus on figure 4.2, the state-space equations of the non-linear system can be derived as follows :

$$\dot{\mathbf{V}} = \mathbf{A}\mathbf{V} + \mathbf{B}V_{in} + \mathbf{f}(\mathbf{V}, V_{in}) \tag{4.1}$$

The order of the state voltages in vector **V** is in accordance with the node-voltages in figure 4.2. The output equation in its most general form is :

$$v_{out} = \mathbf{C}V + \mathbf{D}V_{in} + \mathbf{g}(V, V_{in}) \tag{4.2}$$

A calculation of the state-space matrices in (4.1) and (4.2) gives :

$$\mathbf{A} = \left[\begin{array}{c|c} \overbrace{\mathbf{F}fil}^{m} & \overbrace{\mathbf{0}}^{n} \\ \hline \mathbf{F}fb & \mathbf{F}int \end{array} \right] \begin{array}{l} \}m \\ \}n \end{array} \tag{4.3}$$

with:

$$\mathbf{F}fil =$$

$$\begin{bmatrix}
-2\pi f_{fil-1} & 0 & \cdots & \cdots & \cdots & 0 \\
2\pi f_{fil-2} & -2\pi f_{fil-2} & 0 & \cdots & \cdots & 0 \\
0 & 2\pi f_{fil-3} & -2\pi f_{fil-3} & 0 & \cdots & 0 \\
\vdots & \ddots & \ddots & \ddots & \ddots & \vdots \\
0 & \cdots & \cdots & \cdots & 2\pi f_{fil-m} & -2\pi f_{fil-m}
\end{bmatrix} \tag{4.4}$$

$$\mathbf{F}int =$$

$$\begin{bmatrix}
-\frac{2\pi f_{int-1}}{A_0} & 0 & \cdots & \cdots & \cdots & 0 \\
2\pi f_{int-2} & -\frac{2\pi f_{int-2}}{A_0} & 0 & \cdots & \cdots & 0 \\
0 & 2\pi f_{int-3} & -\frac{2\pi f_{int-3}}{A_0} & 0 & \cdots & 0 \\
\vdots & \ddots & \ddots & \ddots & \ddots & \vdots \\
0 & \cdots & \cdots & \cdots & 2\pi f_{int-n} & -\frac{2\pi f_{int-n}}{A_0}
\end{bmatrix} \tag{4.5}$$

$$\mathbf{F}fb =$$

$$\begin{bmatrix}
0 & \cdots & 0 & -\beta_1 2\pi f_{int-1} \\
0 & \cdots & 0 & -\beta_2 2\pi f_{int-2} \\
\vdots & \vdots & \vdots & \vdots \\
0 & \cdots & 0 & -\beta_n 2\pi f_{int-n}
\end{bmatrix} \tag{4.6}$$

The **B**-matrix and the non-linear **f**(.) function are dependent on the order of the SOPA. In the zeroth order case, these functions become :

$$\mathbf{B} = 0 \tag{4.7}$$

$$\mathbf{f}(V, V_{in}) = \begin{bmatrix} 2\pi f_{fil-1}\frac{V_{DD}}{2}\tanh(A_c(V_{in} - V_m)) \\ 0 \\ \vdots \\ 0 \end{bmatrix} \tag{4.8}$$

The comparator is mathematically modelled as a hyperbolic tangent function with a gain A. When A goes to infinity a *signum*-function, being an ideal comparator in practice, is reached. For a higher order SOPA, the **B**-matrix and the non-linear $\mathbf{f}(.)$ function becomes :

$$
\mathbf{B} = \begin{bmatrix} 0 \\ \vdots \\ 0 \\ 2\pi f_{fint-1} \\ 0 \\ \vdots \\ 0 \end{bmatrix} \begin{matrix} \left.\vphantom{\begin{matrix}0\\ \vdots\\ 0\end{matrix}}\right\} m \\ \\ \left.\vphantom{\begin{matrix}0\\ \vdots\\ 0\end{matrix}}\right\} n \end{matrix} \tag{4.9}
$$

$$
\mathbf{f}(\mathbf{V}, V_{in}) = \begin{bmatrix} 2\pi f_{fil-1} \frac{V_{DD}}{2} \tanh\left(A_c V_{m+n}\right) \\ 0 \\ \vdots \\ 0 \end{bmatrix} \tag{4.10}
$$

Since the output is the output of the non-linear element the **C** and **D** - matrices are both **0**. The output-function (4.2) thus becomes :

$$
v_{out} = \begin{cases} \frac{V_{DD}}{2} \tanh\left(A_c(V_{in} - \beta_1 V_m)\right) & \text{if} \quad n = 0 \\ \frac{V_{DD}}{2} \tanh\left(A_c V_{m+n}\right) & \text{if} \quad n > 0 \end{cases} \tag{4.11}
$$

1.3 Numerical Verification

The numerical verification of the models further developed in this chapter has been done using the ODEPACK *lsode*-routine [Hindmarsh, 1983] compiled in the OCTAVE numerical computation program [Eaton, 2002]. Routines have been implemented in the OCTAVE-language to automatically generate the state-space equations for basic SOPA parameters. Different input signals can be applied to the SOPA in the numerical simulator. Most common signals are : a DC-voltage, a sine wave and different DMT-signals mimicking ADSL or VDSL downstream signals. The lsode differential equation solver calculates the time-domain waveforms on the state nodes. Post-processing is done using a Fast Fourier Transform (FFT) in OCTAVE. Table 4.1 show the default values for the numerical simulator. All examples in this chapter use these values, except otherwise noted.

Two examples of the numerical simulation of a zeroth order SOPA are shown in figure 4.3. The waveforms are taken from node V_m as defined in figure 4.2. Although both simulations are done on the same SOPA model, the

Table 4.1: Default parameters of the numerical simulator

	parameter	value
Simulator		
	algorithm	lsode
	start time	0 s
	step time	1 ns
	stop time	10 ms
	initial values	all states = 0
		except V_1=0.1
	Base resistance for dBm	1 kΩ
	FFT-window	Blackman-Harris
Signal		
	waveform	sine wave
	amplitude	0.1
	frequency	1 MHz
SOPA-parameter		
	order	0
	loop filter order	3
	loop filter cut-off freq.	10 MHz
	integrator unit gain freq.	1 MHz
	supply voltage V_{DD}	3.3 V
	comparator gain	10^6
	β_i	1
	coupling factor α	0.25

results show large differences due to aliasing of the higher order harmonics of the limit cycle oscillation. While the simulation in figure 4.3(a) uses the default step-time, figure 4.3(b) has a reduced step-time of 10 ns. This effect can be predicted as follows : if we assume that the output is a perfect square wave with amplitude $V_{DD}/2$, the energy content of the first harmonic equals :

$$P_{lc} = \left(\frac{2}{\pi} \frac{V_{DD}}{\sqrt{R_{load}}} \right)^2 \tag{4.12}$$

Evaluating this expression for the default values of table 4.1 gives a square wave power of -8 dBm. Since in the ideal mathematical model, the only noise-source are truncation errors in the simulator, the noise floor can be as low as the IEEE relative floating point precision 2.2204e-16. If different distortion phenomena need to be studied without interference due to aliasing, the output square wave need to be sampled at 35.6 times the limit cycle frequency since

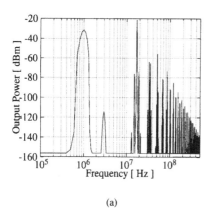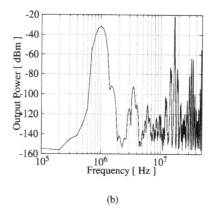

(a)   (b)

Figure 4.3: Two numerical simulation with the same SOPA parameters but different numerical step-time

the Fourier terms decay with 20 dB per decade. The stop-time is determined by the lowest signal frequency that needs to be observable and the roll-off of the window used by the FFT. As a rule of thumb stop-time= $10/f_{sig}$ is used. Since the number of output points equals stop-time/step-time, the simulation of a complete SOPA will be very time and memory consuming. To alleviate these problems, all further simulations are done on the output node of the loop filter for the decay of the loop filter adds to the 20 dB per decade of the square wave. For instance, if the loop filter's order is 3, the decay equals 80 dB per decade and the step-time is reduced to 10 times the limit cycle frequency. In this way, simulation problems are heavily relaxed.

From these considerations it becomes clear that an analytical model is necessary to scan the design space.

2. Zeroth order SOPAs

The discussion of the working principle of the SOPA power amplifier is started with the zeroth order case for its higher simplicity. First the autonomous system (i.e. when no input signal is applied), in a next subsection the forced system equations are derived.

2.1 Limit cycle Oscillation

2.1.1 Analytical determination

Figure 4.4 shows the phase plane trajectories of two numerical simulation of a zeroth order SOPA with a third order loop filter. The loop filter in this

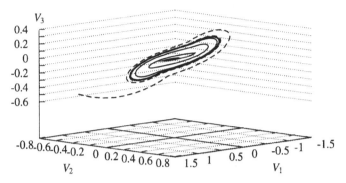

Figure 4.4: Phase plane representation of a zeroth order SOPA power amplifier with a third order loop filter

simulation has a cut-off frequency of 10 MHz and is a cascade of three first order filters with the same cut-off frequency, as in (4.3) and (4.4). Two distinct initial conditions were chosen : $V_0 = 0.1 \ 0 \ 0$ and $V_0 = 1 \ -0.6 \ 0.5$. This to point out a trajectory near the origin and one more at the border of the state space. The two trajectories clearly converge towards a limit cycle. The SOPA thus will self-oscillate when no input signal is applied. The description of this behaviour will be the subject of this subsection.

To calculate this limit cycling behaviour a describing function analysis [Gelb and Vander Velde, 1968] has been performed. The comparator is modelled as (see section 3.2.4.4):

$$N_A(A) = \frac{2V_{DD}}{A\pi} \tag{4.13}$$

If take the cut-off frequencies of the loop filters are taken the same for every section, i.e. $f_{fil-1} = f_{fil-2} = \cdots = f_{fil-n} = f_{fil}$ in (4.4), than the loop filter can be modelled in the frequency domain as :

$$L_f(s) = \left(\frac{2\pi f_{fil}}{s + 2\pi f_{fil}} \right)^n \tag{4.14}$$

The limit cycle frequency f_{lc} and amplitude A can be calculated by solving the Barkhausen criterion [Barkhausen, 1935] : $L_f(s)N_A(A) + 1 = 0$. This complex equation can be split into two real equations : a phase-balance and an amplitude-balance.

$$\mathcal{R}e\left(L_f(s)N_A(A)\right) = -1 \tag{4.15}$$

$$\mathcal{I}m\left(L_f(s)N_A(A)\right) = 0 \tag{4.16}$$

$$2\frac{V_{DD}}{\pi A} \left(4\frac{\pi^2 f_{fil}^2}{4\pi^2 f_{fil}^2 + \omega^2} \right)^{n/2} \cos(n \ \arctan(\frac{\omega}{2\pi \ f_{fil}})) = -1 \tag{4.17}$$

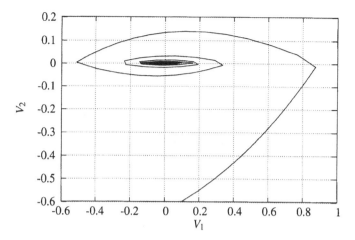

Figure 4.5: Phase plane trajectory of a zeroth order SOPA using a loop filter with order = 2, obtained using numerical simulation.

$$2\frac{V_{DD}}{\pi A}\left(4\frac{\pi^2 f_{fil}^2}{4\pi^2 f_{fil}^2 + w^2}\right)^{n/2}\sin(n\,\arctan(\frac{w}{2\pi\,f_{fil}})) = 0 \qquad (4.18)$$

The limit cycle frequency can be calculated from equation (4.18)

$$f_{LC} = f_{fil}\tan\left(\frac{\pi}{n}\right) \qquad (4.19)$$

From equation (4.19), we can derive that a necessary condition for limit cycle oscillation is that the order of the loop filter should be greater than 2, since no valid solution can be found. In case of $n = 2$ the phase plane trajectories will be damped spirals as shown in figure 4.5. The limit cycle amplitude can be found by filling in equation (4.19) in (4.17).

$$A = \frac{2V_{DD}}{\pi}\cos^n\left(\frac{\pi}{n}\right) \qquad (4.20)$$

2.1.2 Graphical representation

This solution can be graphically presented by the modified Nyquist plot as explained in section 3.2.4.3 . The crossing points of the loop filters Nyquist curve $G(j\omega)$ and $\frac{-1}{N(A)}$ provide possible limit cycle oscillations. Only the ones that are an even times encircled by the $G(j\omega)$ curve, are stable limit cycles.

Solution for a zeroth order SOPA with a loop filter's order of 2 (a), 3 (b), 8 (c) and 12 (d) are drawn in figure 4.6. Stable limit cycle solutions are marked with a cross, unstable with a circle. From these figures the following conclusions can be drawn:

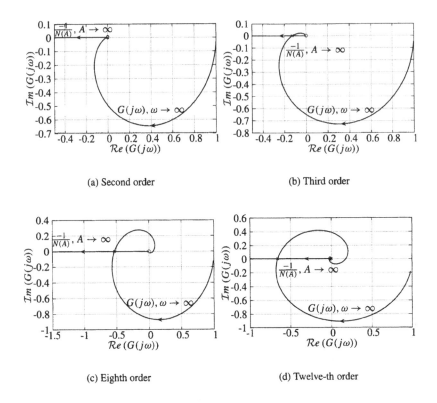

(a) Second order   (b) Third order

(c) Eighth order   (d) Twelve-th order

Figure 4.6: Modified Nyquist chart for a zeroth order SOPA with a different order loop filters. The scales of the real and imaginary axes are taken unequal to stress the crossing point.

- Stable limit cycles only exist for filter orders greater than two, otherwise no crossing of the negative x-axis is possible. This is equal to the phase balance of the Barkhausen criterion.

- There will always be at least one stable solution for a filter order greater than 2. The solution with the lowest frequency is given by (4.19).

- Filter orders higher than 12 will generate more stable solutions. In theory, the system can oscillate in either mode depending on the initial conditions. In practice, these solutions will be so high frequent that they will not occur, and only the solutions of (4.19) will be observed.

Also from these figures, it can be determined that the simplification $f_{fil-1} = f_{fil-2} = \cdots = f_{fil-n} = f_{fil}$ makes sense, from a design point-of-view since:

- The limit cycle frequency will be determined by the lowest filter poles.

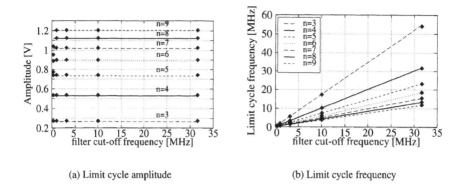

(a) Limit cycle amplitude

(b) Limit cycle frequency

Figure 4.7: Numerical verification (diamonds) of the results obtained using the describing function analysis (solid line)

- The feedback will be used in the forced oscillation case to lower the distortion. The lowest frequency pole has to be higher than the systems bandwidth. Since it is preferred that the mean switching frequency does not exceed the signal bandwidth too much, the filter poles will be close to each other.

2.1.3 Numerical verifications

The approximations that are the base of the describing function technique do not always hold. For non-linear design, these approximations should always be checked by numerical simulations.

Figure 4.7 shows the comparison between numerical calculations, indicated by a diamond and the results from the describing function analysis (4.19) and (4.20). The limit cycle amplitude and frequency are calculated for different values of the loop filter's cut-off frequency and order. A perfect match can be observed. The small deviations for low cut-off frequencies are due to the numerical inaccuracy of the FFT. The describing function analysis accuracy improves with increasing filter order. This was expected, since the filter hypothesis [Gelb and Vander Velde, 1968] is more fulfilled if higher harmonics are filtered out. So for any working SOPA system, the describing function analysis holds and can thus be used in the design process.

2.2 Coupled System Equations

2.2.1 Resistive Coupling

As depicted in figure 4.1, two SOPA building blocks are coupled via the load. If we assume a resistive loadcoupling R_L and a non-ideal output buffer

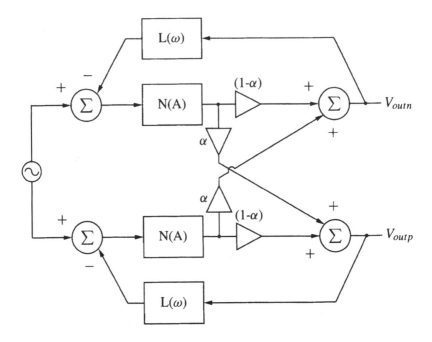

Figure 4.8: Block schematic of a coupled differential SOPA amplifier.

with output resistance r_{out}, the coupling factor α can be calculated as follows:

$$\alpha = \frac{r_{out}}{2r_{out} + R_L} \qquad (4.21)$$

The coupling factor α is in case of a resistive load always between 0 and 0.5. The complete block schematic is given in figure 4.8. The coupling is modelled by cross-coupled amplification that splits the signal in a forward coupling $(1 - \alpha)$ and a cross-coupling amplification α. If the load resistance goes to infinity, the coupling diminishes. The comparator is modelled by its describing function $N(A)$, while the linear loop-filter is represented by its transfer function $L(\omega)$.

The block schematic of figure 4.8 can be simplified by simple block algebra to the schematic of figure 4.9. From this diagram the loop gain can be calculated easily. The Barkhausen criterion is thus given by :

$$\frac{\alpha^2(N(A)L(\omega))^2}{(1 + (1 - \alpha)N(A)L(\omega))^2} = 1 \qquad (4.22)$$

This equation can be further simplified to :

$$(1 + N(A)L(\omega))\,(1 + (1 - 2\alpha)N(A)L(\omega)) = 0 \qquad (4.23)$$

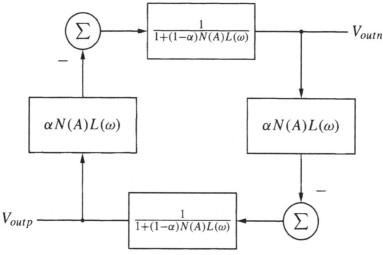

Figure 4.9: Simplified block schematic of figure 4.8

This can be interpreted as follows : there exist 2 oscillation modes in the coupled system.

1. One is solution of equation $(1 + N(A)L(\omega))$ and represents an in phase oscillation of the two coupled SOPA amplifiers [Lindgren, 1964]. This equation is the same as the single SOPA limit cycle calculation (4.16). This is logical since there does not flow current, thus no information, through the load resistance if the SOPAs oscillate in phase. In this mode, the two SOPAs oscillate as if they were not coupled. This in phase oscillation has thus the same limit cycle amplitude and frequency as the single SOPA-case.

$$\omega_1 = \omega_c \tan\left(\frac{\pi}{n}\right) \tag{4.24}$$

$$A_1 = \frac{2\,V_{DD}}{\pi}\cos^n\left(\frac{\pi}{n}\right) \tag{4.25}$$

2. The other solution are roots of the equation $(1 + (1 - 2\alpha)N(A)L(\omega))$. This solution represents the case where the 2 SOPAs oscillate in counter-phase. For a resistive coupling, α is a real number and the roots of this equation can be directly derived from the single SOPA-case by substituting $(1-\alpha)N(A)$ by $N'(A)$. The solutions thus become :

$$\omega_2 = \omega_c \tan\left(\frac{\pi}{n}\right) \tag{4.26}$$

$$A_2 = \frac{2\,V_{DD}}{\pi}(1 - 2\alpha)\cos^n\left(\frac{\pi}{n}\right) \tag{4.27}$$

A stability analysis has to be performed to determine which mode/modes will occur in the coupled system. From a line driver performance point of view, it is preferred that the in-phase oscillatory mode is stable and the counter-phase mode is unstable since :

- When both oscillatory modes are unstable, the self-oscillations are mutually killed. Since the self-oscillation is necessary to dither the input signal and is thus required for proper modulation, this mode is mostly unwanted.

- The in-phase mode is preferred for two important reasons :

    1 When no signal is applied, the two SOPAs will oscillate in-phase. The oscillation is thus common mode for the differential load. The power consumption of the line driver thus is heavily reduced when no input signal is applied and this without shutting down the power amplifier.

    2 The oscillator pulling of the two SOPAs filters out the mean switching frequency. Their is no extra filtering necessary in the output power path to remove the limit cycle.

Furthermore, when comparing the limit cycle amplitudes A_1 and A_2 from (4.24) and (4.26), the counter-phase amplitude is lower than the in-phase limit cycle amplitude. As will be explained in section 4.2.3, where the driven SOPA is handled, the limit cycle will act as a natural dither signal. By lowering the dither level, the linearity will decrease.

- If the two oscillatory modes are stable, the single-input describing function analysis is still valid since from equations (4.24) and (4.26), it is shown that the oscillation frequency is the same in both modes. The system will thus oscillate in the mode where the stability gain is the highest (see appendix A). The oscillation can start up in either mode, but a small perturbation will lead it to the dominant mode, i.e. the one with the highest loop gain.

The derivation of the stability criterion is subject of appendix A. The resulting stability criteria become for the in-phase limit cycle oscillation :

$$\frac{n\pi \cos^{(2-n)}\left(\frac{\pi}{n}\right)}{2\alpha^2 V_{DD}\omega_c} > 0 \qquad (4.28)$$

This is always fulfilled since $n > 3$. For the counter-phase component the stability criterion is :

$$\frac{n(1-2\alpha)\pi \cos^{(2-n)}\left(\frac{\pi}{n}\right)}{2\alpha^2 V_{DD}\omega_c} > 0 \qquad (4.29)$$

This criterion is not fulfilled if $\alpha \geq 0.5$. For the normal operation $0 < \alpha < 0.5$ both oscillations could be stable. To determine which oscillation will be present, the TSIDF representation of the system has to be calculated. The transfer function of a small disturbing in-phase oscillating signal to the system if the system is oscillating in counter phase mode, is calculated (TF_{common}). The same is done with a small counter-phase disturbing signal on a in-phase oscillating system ($TF_{counter}$). The mode with the highest Dual Input Describing Function (DIDF) gain will be the mode occurring in a physical system. The resulting small signal transfer functions for the common mode case (TF_{common}) and the counter phase mode ($TF_{counter}$) become :

$$TF_{common} = \left| \frac{\alpha^2 (N_{A_2}(A_1, A_2)L(\omega_0))^2}{(1 + (1 - \alpha)N_{A_2}(A_1, A_2)L(\omega_0))^2} \right| \tag{4.30}$$

$$TF_{counter} = \left| \frac{\alpha^2 (N_{A_1}(A_1, A_2)L(\omega_1))^2}{(1 + (1 - \alpha)N_{A_1}(A_1, A_2)L(\omega_1))^2} \right| \tag{4.31}$$

in which

$$N_{A_1}(A_1, A_2) \approx \frac{V_{DD}}{\pi A_2} \tag{4.32}$$

$$N_{A_2}(A_1, A_2) \approx \frac{V_{DD}}{\pi A_1} \tag{4.33}$$

We define also the excess common mode gain to be :

$$ECMG = 10 \log \left(\frac{TF_{common}}{TF_{counter}} \right) \tag{4.34}$$

The Excess Common Mode Gain (ECMG) is positive when the common mode oscillation has the highest gain and is thus the one occurring in an implemented system. It also denotes the speed by which the system from a counter mode impulse recovers to the common mode oscillation. Filling (4.24) and (4.26) in (4.30) and (4.31) will generate the following gains :

$$TF_{common} = \frac{\alpha^2}{(1 - 3\alpha)^2} \tag{4.35}$$

$$TF_{counter} = \frac{\alpha^2}{(1 + \alpha)^2} \tag{4.36}$$

The condition for the system to oscillate in phase thus becomes :

$$\frac{TF_{common}}{TF_{counter}} = \frac{(1 + \alpha)^2}{(1 - 3\alpha)^2} > 1 \tag{4.37}$$

$$\Rightarrow \alpha < 1 \tag{4.38}$$

This is fulfilled for α values between 0 and 1. Since $0 < \alpha < 0.5$, the system will always oscillate in phase. The outcome of this analysis can be easily verified by physical considerations :

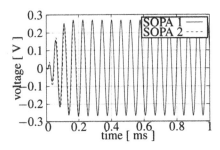

(a) All initial condition =0. Small perturbation on 1 node.

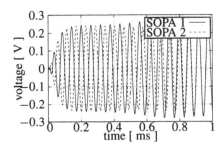

(b) All initial condition =0. Small differential perturbation on 2 nodes

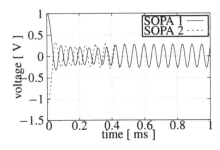

(c) Differential initial conditions.

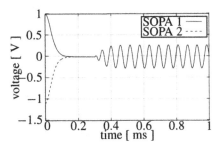

(d) Differential initial conditions. Coupling factor $\alpha = 0.5$

Figure 4.10: Numerical simulations of two coupled SOPAs

- If the two SOPAs do not oscillate in-phase, a considerable energy is consumed. This lowers the energy-content in the loop filter, since via r_{out}, the delivery of energy to the system by the comparator is limited. This energy-consideration is similar to a Lyapunov stability analysis [Slotine and Li, 1991].

- If $\alpha = 0.5$, $R_L = 0$. The only possible solution is the in-phase oscillation, since in this case the output nodes are connected.

- If α exceeds 0.5, a negative resistance is connected between the output nodes. Only the in-phase limit cycle is stable. Due to the negative resistance more energy is inserted in the SOPA-loops. The mode with the highest amplitude will thus be the only excited one.

To illustrate these findings, several numerical simulations are shown in figure 4.10. The small disturbances are simulated by adding a small quantity to one or more of the initial conditions. Except for simulation 4.10(d), all simulations were done with rather low coupling ($\alpha < 0.1$), to have a better visualisation of the coupling effects.

In figure 4.10(a) all initial conditions are set to zero. The system is thus in a meta-stable state. By setting a small initial voltage on one of the filter nodes, the system will diverge from this state to the most likely stable oscillation. The system will quasi instantaneously converge to the stable in phase oscillation.

In figure 4.10(b) , all initial condition were also set to zero as in the previous case, but a differential initial disturbance, i.e. an opposite small disturbance of the initial condition on two corresponding nodes of both SOPAs, has been applied. This differential excitation should trigger the counter-phase oscillation mode. In the start-up phase this can be clearly observed. The systems start to oscillate in counter-phase mode with the same oscillation frequency, but lower amplitude as it has been predicted by (4.26). This mode however, will not persist due to the higher gain of the in phase mode. Small perturbations, like noise (in these models represented by the limited numerical inaccuracy), will drive the system out of this stable mode towards in phase oscillation.

In figure 4.10(c) , the system is started with a maximum differential excitation. The system starts oscillating in counter phase, but the two SOPAs start diverging due to the difference in loop gain of the two modes.

In figure 4.10(d) , the same conditions as in the previous case were used. The coupling factor α however was set to 0.5. As been predicted by (4.29), the differential mode is unstable. No differential oscillation can be started. The initial conditions exponentially decrease and the in phase mode is started after this start-up behaviour.

2.2.2 Non-Resistive Coupling

Since real lines are never purely resistive at the limit cycle frequencies, the attraction of both SOPA amplifiers should be investigated for a non-resistive load as well. The oscillation modes were investigated using a line model which has an important inductive impedance (figure 4.11(a)) and a capacitive one (figure 4.12(a)). For the inductive line model an inductor of 0.6 mH/km has been taken, while for the capacitor has a value of 49 nF/km, with a maximum line length of 6 km. The DC resistance of the line has been set to 100 Ω. Next to the line-models, a bode plot of the coupling factor α for increasing frequency ω and inductance L (figure 4.11(b)), respectively frequency ω and capacitance C (figure 4.12(b)). The characteristics are expressed in function of the DC-coupling factor α_0, on itself being a function of the output resistance

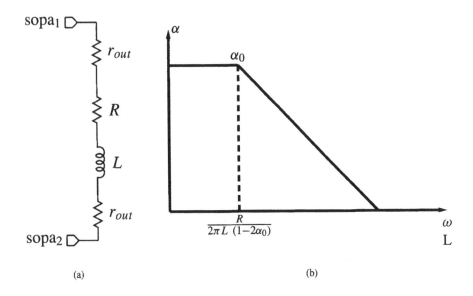

Figure 4.11: Line model for a inductive line impedance and the corresponding bode plot for its coupling factor

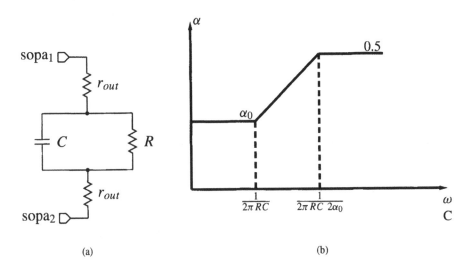

Figure 4.12: Line model for a capacitive line impedance and the corresponding bode plot for its coupling factor

and the (fixed) transformed characteristic line impedance, and the maximal capacitance, resp. inductance values. A closer look at the bode plot of the inductive line model reveal the presence of one pole which is dependent on α_0. The coupling factor thus rolls off with increasing frequencies and the position of the corner frequency is proportional with α_0. For the capacitive case, a fixed zero is observed at $1/(2\pi\ RC)$ and a pole at a factor $1/(2\alpha_0)$ higher. Above this pole frequency the coupling is at a maximal value of 0.5 . Notice that the pole/zero gap and thus also the phase shift between the two SOPAs narrows with higher α_0 values.

The common mode oscillation frequency and amplitude is independent on the nature of coupling, since the coupling factor disappears in the expression for the loop gain (4.23). The expressions for the limit cycle frequency and amplitude are the same as in (4.24).

To calculate the counter mode, the frequency behaviour of the $(1-2\alpha)$ has been plotted for the capacitive (figure 4.15) and the inductive case (figure 4.13). For the inductive line model, this $(1-2\alpha)$ factor can be calculated as follows :

$$(1-2\alpha) \quad = \quad 1 - 2\frac{r_{out}}{2\,r_{out} + R + sL} = 1 - 2\frac{\alpha_0}{1 + j\frac{\omega}{\omega_L}} \tag{4.39}$$

$$= \quad (1-2\alpha_0)\frac{1 + j\frac{\omega}{\omega_L(1-2\alpha_0)}}{1 + j\frac{\omega}{\omega_L}} \tag{4.40}$$

with :

$$\alpha_0 = \frac{r_{out}}{2r_{out}+R} \tag{4.41}$$

$$\omega_L = \frac{R}{L(1-2\alpha_0)} \tag{4.42}$$

For the capacitive line model, the calculations yield the following expressions :

$$(1-2\alpha) \quad = \quad 1 - 2\frac{r_{out}}{2\,r_{out} + \frac{R}{1+sCR}} = 1 - 2\frac{\alpha_0(1 + j\frac{\omega}{\omega_C})}{1 + j\frac{i\omega}{\omega_C/2\alpha_0}} \tag{4.43}$$

$$= \quad (1-2\alpha_0)\frac{1}{1 + j\frac{\omega}{\omega_C/2\alpha_0}} \tag{4.44}$$

with :

$$\alpha_0 = \frac{r_{out}}{2r_{out}+R} \tag{4.45}$$

$$\omega_C = \frac{1}{RC} \tag{4.46}$$

The α_0 factor can be coupled to the output efficiency ϵ since the losses in first order can be attributed to the output resistance.

$$\epsilon = \frac{R}{2r_{out} + R} \tag{4.47}$$

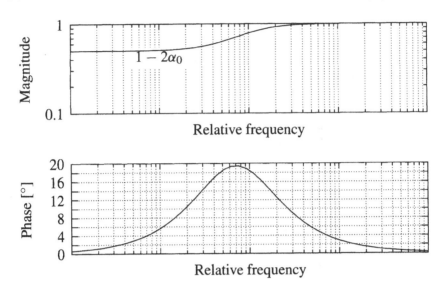

Figure 4.13: Bode plot of the $(1 - 2\alpha)$ factor used to calculate the counter oscillatory mode frequency and amplitude, for an inductive line model. $\alpha_0 = 0.25$

Substituting (4.47) in (4.45) gives the practical range of α_0 :

$$\alpha_0 = \frac{1 - \epsilon}{2} \qquad\qquad (4.48)$$

The α_0-factor should thus be as low as possible to provide a high efficiency. However, if active line termination is used as has been introduced in section 2.4.2 , the r_{out} can be made equal to $R/2$ without degrading the efficiency too much. In this case, the α_0 factor will be 0.25. In the following, the resistive coupling factor is assumed to be between 0 and 0.3. This will provide enough safety margin for the conclusions to hold in practical designs.

Taking into account that the $1 - 2\alpha$ factor in the counter mode Barkhausen criterion acts as an extra loop filter in the case of a complex load, some qualitative remarks can be drawn from the observations of figure 4.13 and figure 4.15 :

The inductive line model will lead to lower coupling when the line gets more inductive. From figure 4.13, the limit cycle frequency and amplitude can be estimated. Since the $(1 - 2\alpha)$ filter consists of an early zero situated at $R/(2\pi L)$, followed by a pole at higher frequencies dependent on α_0, the inductive line has no influence at low α_0 factors. For higher α_0-factors the counter mode limit cycle frequency will be slightly higher due to the positive phase shift of the $(1 - 2\alpha)$ filter than for the common mode case. The

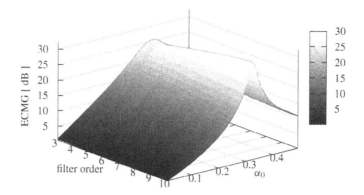

Figure 4.14: Representation of the calculated excessive common mode gain for an inductive line, calculated as function of α_0 and the filter order n.

gain of this extra loop filter is always lower than 1, and since the loop transfer function is low-pass for the filter criterion to hold, it is to be assumed that for the inductive model the ECMG will always be positive.

To determine which oscillation mode will occur in a physical system, the ECMG, as it has been defined in (4.34), is calculated. For this the phase influence of the $(1 - 2\alpha)$ filter is taken into account in the calculation of the counter mode limit cycle frequency. The counter mode limit cycle frequency (ω_2) is thus the solution of the following equation :

$$m \ \arctan\left(\frac{\omega_2}{\omega_{fil}}\right) + \arctan\left(\frac{\omega_2}{\omega_L}\right) - \arctan\left(\frac{\omega_2}{\omega_L(1 - 2\alpha_0)}\right) = \pi \quad (4.49)$$

Filling in ω_2 in the amplitude criterion and taking the α characteristic of figure 4.11(b) into account leads to a straightforward expression for the ECMG. The results of this calculation is presented in figure 4.14.

From figure 4.14, the stated observations were confirmed. For low α_0 ranges the influence of the inductance is unimportant. The ECMG increases from $\alpha_0 = 0$, meaning no coupling, towards a maximum at about $\alpha_0 = 1/3$. Which is in accordance with (4.37). For higher values of α_0 the phase shift connected with the inductive line characteristic becomes more pronounced and a higher value for the ECMG is to be observed compared with the predictions of the resistive model (4.37). But these high resistive values are too high for a real implementation, since the output efficiency will be too low due to a over-dimensioned output resistance. However, also for a more inductive line, a coupling factor around 0.25, meaning the introduction of a line termination network will also aid the coupling.

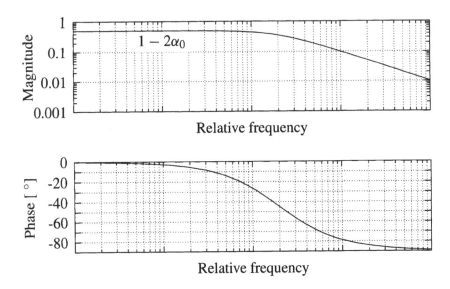

Figure 4.15: Bode plot of the $(1 - 2\alpha)$ factor used to calculate the counter oscillatory mode frequency and amplitude, for a capacitive line model. ($\alpha_0 = 0.25$)

When going to higher filter orders the ECMG slightly improves, but not significantly to constrain the filter order specifications.

In the capacitive case , the contribution of the $(1 - 2\alpha)$ factor in the counter mode oscillation transfer function acts like a regular first order filter with a DC gain of $(1 - 2\alpha_0)$. The corner frequency is inversely proportional with the DC-coupling factor α_0. The bode plot of this extra filter is depicted in figure 4.15. As long as the coupling factor is low enough for this corner frequency to lie within the frequency band of interest, the $(1 - 2\alpha)$ coupling factor behaves as an extra loop filter, increasing the loop filters order by one. This means that for coupled SOPA systems with a loop filter of two, the coupled system will always oscillate in counter phase, since the common mode oscillation is unstable. This observation puts a clear restriction on the applicability of a SOPA, being that the order of the loop filter is three or higher.

Due to the extra phase shift introduced by the coupling, the limit cycle oscillation will thus be at lower frequencies than the common mode oscillation frequency. The limit cycle amplitude will be around the values found in the common mode case for the increased filtering compensates the amplitude lower filtering due to the decrease in oscillation frequency. The consequence of this could be a decrease in ECMG since the counter mode

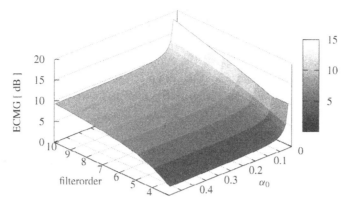

Figure 4.16: Representation of the calculated worst-case excessive common mode gain for a capacitive line, calculated as function of α_0 and the filter order n.

limit cycle oscillation at lower frequency thus has a higher loop gain if only the filters are considered.

Another important observation can be made from the bode plot of the coupling factor, as shown in figure 4.12(b). Due to the capacitive load, the coupling factor increases towards a maximum of 0.5. Since the increase of coupling factor normally would mean a higher ECMG, the common mode should be dominant.

The ECMG is calculated to determine the dominant effect. This could again be done straightforwardly since it only takes the addition of an extra filter block to calculate the counter mode frequency and amplitude. Filling these values in the loop transfer function together with the coupling factor characteristic of figure 4.12(b) leads to an explicit[1] expression for the ECMG. The results for a capacitive line are given in figure 4.16.

From figure 4.16, the following observations can be drawn : the ECMG increases drastically for low DC coupling factors. This is due to the fact that the effective coupling factor is always higher than the α_0 value. The common mode oscillation is always stable but the ECMG is lower than in the inductive and the resistive case. For high filter orders this effect diminishes due to the decreased importance of the addition of one extra order by the $(1 - 2\alpha)$ filter block compared with the installed loop filter.

---

[1]With an explicit expression, a mathematical expression is meant which evaluates directly to the wanted parameter in a form $y = f(\mathbf{x})$. We will use this term in the next of this chapter on those places where an explicit expression can be derived which is easy evaluated but which is too large, mostly due to the high numbers of parameters, to be printed in the text.

2.2.3   Concluding remarks on coupling

By the coupling of two SOPA amplifiers by the load, only two oscillation modes between two SOPAs exist: the common oscillation mode in which the two SOPAs oscillate in-phase and the counter oscillation mode where a 180° phase shift between the limit cycle oscillations is maintained. The common mode oscillation mode is preferable since :

- the common mode oscillation has intrinsically a higher efficiency, due to the fact that the limit cycle oscillation is not transferred to the load.

- the limit cycle frequency and amplitude is independent from the load conditions.

The Excess Common Mode Gain (ECMG) is a measure for the stability of the common mode signal and denotes the speed by which a counter mode initial condition fades out towards a common mode oscillation. The ECMG was calculated and evaluated for a resistive, inductive and capacitive load, spanning the extremes of a real line. In all cases the common mode oscillation is dominant. However if the SOPA is used as a one-way power amplifier like for instance as an audio driver and thus no impedance matching is demanded, or if the impedance matching is performed at the line side of the transformer, the coupled SOPA has a low coupling factor. The addition of a parallel capacitor is in those cases beneficial to increase the ECMG. For all line models, the ECMG is increasing for increasing α_0 factors up to $1/3$. So, the inclusion of active impedance synthesis should improve the coupling properties of the differential SOPA system.

2.3   Forced System Oscillation

2.3.1   Dithering effect of the limit cycle oscillation

In this section the response of a zeroth order SOPA to an external sinusoidal input signal is examined. It is assumed that the zeroth order is designed in such a way that the common mode oscillation occurs as it has been explained in the previous sections. Therefor we can simplify the forced oscillation calculations in first order to the uncoupled system of system equations (4.1).

To calculate the behaviour of the SOPA when an external signal is applied, i.e. the forced system response, the TSIDF representation of the comparator is used. For an ideal comparator, this gives for the gain of the input signal with an amplitude B at the comparators input, in the presence of a limit cycle oscillation with amplitude A :

$$N_B(A, B) \;=\; \frac{V_{DD}}{\pi B}\left(\frac{B}{A}\right) {}_2F_1\left(\frac{1}{2}, \frac{1}{2}; 2; \left(\frac{B}{A}\right)^2\right)$$

$$\approx \; \frac{V_{DD}}{\pi A} = N(A)/2$$
$$\text{when } 0 < B \lll A \tag{4.50}$$

In which ${}_2F_1(a, b, ; c; z)$ denotes the 2-1 hyper-geometric function in the variable z with factors (a, b) and (c). It can be calculated as

$${}_2F_1(a, b, ; c; z) = \sum_{n=0}^{\infty} \frac{(a)_n (b)_n}{(c)_n} \frac{z^n}{n} \tag{4.51}$$

$$\text{with } (a)_n = \frac{\Gamma(x + n)}{\Gamma(x)}$$

Equation 4.49 shows the dithering character of the limit cycle oscillation. Indeed, as long as the amplitude B is smaller than the limit cycle amplitude, the comparators gain will be independent of the signal amplitude B. For small values of B, the comparator will act as a linear gain block. Since factor B is the error signal corresponding with the input signal (V_{in}) under feedback, B can be expressed as :

$$B = \frac{V_{in}}{1 + N(A)/2 \; L(\omega)} \tag{4.52}$$

2.3.2 Dynamic Range Calculation

In order to estimate the in-band distortion, the second term of (4.49) is also taken into account.

$$N_B(A, B) = \frac{V_{DD}}{\pi A} + \frac{V_{DD} B^2}{8\pi A^3} + O\left(\frac{B^4}{A^4}\right) \tag{4.53}$$

Since the second gain term is dependent on the input referred error amplitude B, third order distortion will be generated, since

$$\sin^3(x) = 3/4 \sin(x) - 1/4 \sin(3x) \tag{4.54}$$

This gives for the calculation of the distortion the block schematic as depicted in figure 4.17. In this schematic the main forward contribution to the error-signal B is split from the contribution by the third order gain (upper branch). With bandpass filters the different frequency components are split. The distortion at three times the frequency ω can be considered as a distortion signal which is injected behind the comparator and is dependent on the amplitude of the error signal B at frequency ω. It is to be noted that the describing function for the comparator in the 3ω-branch is given by $V_{DD}/(2\pi\,A)$. The extra $1/2$ factor is due to the fact that this describing function is not a two sinusoid describing function but a three sinusoid. Indeed, the third order harmonic is present in the loop while two other sinusoids are present. The amplitude criterion is fulfilled with respect to both of these oscillations: the third order is smaller than the limit cycle oscillation and obviously also smaller than the first harmonic. The relative amplitude of the signals present in the SOPA decouples the problem in three consecutive and similar steps :

1 The limit cycle oscillation is the largest signal in the system. The calculation of this signal can be done by neglecting the other signals, i.e. the applied input signal and its harmonics. The calculation of this signal can be done by the single sinusoid describing function as it has been proposed in section 4.2.1. This approximation is valid as long as the input signal is sufficiently small. In the wanted working region of the SOPA, this condition is always fulfilled, since a linear modulation is wanted.

2 The second largest signal is the driving signal. Since the limit cycle oscillation is still present in the system, it cannot be neglected. Therefor the TSIDF needs to be used. To improve the mode, the first two terms of the TSIDF are taken, since a third power term will also generate signal at the input frequency (4.54). It is assumed that the limit cycles amplitude and frequency is not influenced by this applied signal. The values from the previous steps are taken unaltered. Another consequence of the discrimination in amplitude is that the gain of the comparator in the loop is almost a half of the gain in the single sinusoid case. In the block scheme, this step is represented by closing the loops with the bandpass filters at ω. The values of the signals at this frequency can be calculated with a high accuracy using these simplifications. So, the error-value B can be determined in this step.

3 In the final step, the distortion is assumed to be injected at the output of the comparator. The value of this distortion signal is given by the previously calculated error-signal B and the second term of the TSIDF approximation. This signal is suppressed by the loop gain which is represented by closing the loop through the bandpass filters at 3ω. Since there are two more signals present in the loop which cannot be neglected, the TSIDF is no longer valid. Therefor the gain of the comparator needs to be calculated using the three sinusoid describing function. The aim is to create a linear amplifier, so the third order distortion is by construction much lower than the limit cycle oscillation and the wanted signal. Also the wanted signal can be neglected with respect to the limit cycle frequency. The three sinusoid describing function can in this way be simplified to one fourth of the single input describing function.

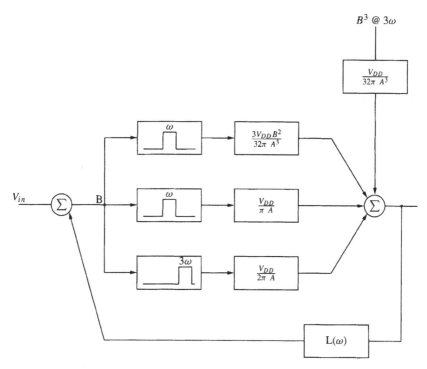

Figure 4.17: Block schematic of the SOPA expanded for the calculation of the third order distortion. The loop is cut for different frequency by the insertion of extra bandpass filters.

Since these three steps are consecutive without iteration, the distortion can be calculated straightforwardly. Note that also higher order harmonics an be calculated by adding further steps.

From the resulting block scheme the distortion can be calculated. Therefor the total error signal B at the input frequency ω needs to be calculated. The system of figure 4.17 needs therefor to be calculated at the input frequency.

$$\frac{V_{DD}}{\pi A} B \, L(\omega) + \frac{3 \, V_{DD}}{32\pi \, A^3} B^3 \, L(\omega) + B = V_{in} \qquad (4.55)$$

This equation can be solved for B.

$$B = \frac{A\pi}{(V_{DD}L(\omega) + \pi A)} V_{in} + O(V_{in}^3) \qquad (4.56)$$

Calculating the loop transfer function for the signal at 3ω gives then :

$$S \times \text{HD3} = \frac{V_{DD}}{32\pi \, A^3 (1 + V_{DD}/(2\pi \, A)L(3\omega))} B^3 \qquad (4.57)$$

with S the signal power at the output:

$$S = \frac{N(A)}{1 + N(A)L(\omega)} V_{in} \tag{4.58}$$

A numerical example. will clarify the use of these formulas. Assume the distortion of a zeroth order SOPA needs to be determined for an input signal with an amplitude of 0.1 V. The parameters of the SOPA are the same as the ones from table 4.1.

Firstly, the limit cycle properties need to be determined. Therefore, the appropriate parameters need to be filled in in (4.19) and (4.20). This gives:

$$f_{lc} = 10 \text{ MHz} \tan\left(\frac{\pi}{3}\right) = 17 \text{ MHz} \tag{4.59}$$

$$A_{lc} = \frac{23.3 \text{ V}}{\pi} \cos^3\left(\frac{\pi}{3}\right) = 0.26 \text{ V} \tag{4.60}$$

Next step is to calculate the error-signal corresponding with the input B at the input of the comparator. Equation 4.56 is limited to its first order term for the hand calculations.

$$L(\omega) = \left(\frac{10 \text{ MHz}}{\sqrt{(10 \text{ MHz})^2 + (1 \text{ MHz})^2}}\right)^3 = 0.985 \tag{4.61}$$

$$B = \frac{0.26 \text{ V}\pi}{0.985 \ 3.3 \text{ V} + 0.26 \text{ V}\pi} 0.1 \text{ V} = 0.02 \text{ V} \tag{4.62}$$

The output signal swing is :

$$S = \frac{2.02}{1 + 2.02 \ 0.985} 0.1 \text{ V} \tag{4.63}$$

Equation 4.57 can be calculated as:

$$S \times HD3 = \frac{3.3}{32\pi (0.26)^3 (1 + 3.3 \ 0.88/(2\pi 0.26)} 0.02^3 = 5.4 \text{ μV} \tag{4.64}$$

The HD3 figure for this configuration can thus be easily estimated to be:

$$HD3 = 20 \log\left(5.4 \ 10^{-6}\right) - 20 \log\left(0.0676\right) \tag{4.65}$$
$$= -105.4 + 23.4 = -82 \text{ dB} \tag{4.66}$$

The complete calculation including the third order term of B in (4.56) is evaluated and compared with numerical simulations in figure 4.18. This distortion can be calculated for different orders of the loop filter. Figure 4.19 shows this calculation for filter orders n from 3 to 10. The distortion only has a weak

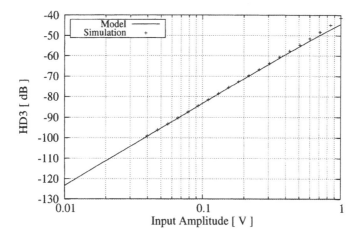

Figure 4.18: Evaluation of the in-band distortion model compared with numerical simulations.

dependence on n. This is due to the combined effect of (4.25) on (4.53) and (4.57):

- The limit cycle amplitude increases with increasing n. The amount of distortion generated by the non-linear gain of (4.53) will decrease with the second power of A.

- The loop gain, however, will drop with increasing n. This is due to a lower linear gain with increasing limit cycle amplitude since the linear gain in (4.53) is inversely proportional with the limit cycle amplitude.

- The loop gain at the third order distortion also decreases for higher filter orders due to the higher roll-off of the loop filter at three times the frequency, i.e. the $L(3\omega)$ factor in (4.57).

If you take the conclusions of section 4.2.2 about the oscillator pulling and the coupling of the SOPA for different filter others into account, a loop filter order of $n = 3$ seems to be the better choice.

To describe the performance of the SOPA amplifier for a real ADSL-system , the MTPR has to be calculated. First of all, the output level of the SOPA is limited by the supply voltage. For ADSL however a signal with a power of 20 dBm needs to be delivered on a 100 Ω line, which is equivalent with an rms-voltage of 3.16 V. This rms-voltage needs to be multiplied with the crest factor to determine the full signal swing. This the becomes 5.6×3.16 V=17.7 V. In order to reach these high voltage levels, a signal transformer is used with a transformer ratio y. An increase in transformer ratio will result in lower distortion, since signal levels at the comparators input are lower as can be observed

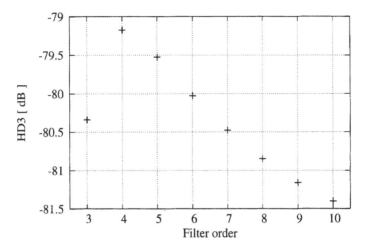

Figure 4.19: Evaluation of the in-band distortion model evaluated for different filter orders n.

from figure 4.18. This however comes at the cost of up-transformed noise. The noise at the output of the SOPA amplifier is multiplied with the transformer ratio. The total noise-level at the output has to be below -100 dBm/Hz which is equivalent with 3.2 $\mu V/Hz^{1/2}$. Analytically this can be expressed as follows: the distortion induced MTPR degradation in function of the rms-voltage σ is calculated in (4.67).

$$\text{MTPR}_{\text{disto}} = 2N \int_0^{\frac{\text{CF} \cdot \sigma}{y}} \frac{2y \ (\text{HD3}(V_{in}))^2 \ e^{\left(-\frac{x^2 y^2}{\sigma^2}\right)}}{\sqrt{2\pi}\sigma} dx \qquad (4.67)$$

In this equation N denotes the number of tones of the DMT-signal, CF denotes the crest factor of the ADSL-modulation, $\text{HD3}(V_{in})$ is the distortion function as it has been calculated in (4.57). It has be shown in section 2.3.2 that a DMT-symbol can be assumed to be a signal with an ideal Gaussian amplitude distribution. This is a good approximation for signals with enough tones. From (4.67) and figure 4.18, it can be concluded that a higher MTPR can be reached when going to higher transformer ratios since the signal levels relax. The maximum achievable MTPR, however, is a combination of the distortion and the noise energy at the output of the amplifier σ_n^2. A good approximation for the upper limit of the achievable MTPR of an almost ideal SOPA amplifier (it is called an almost ideal SOPA since noise is taken into account), can be calculated as :

$$\text{MTPR} = 10 \log \left(\frac{20 \text{ dBm}}{\text{MTPR}_{\text{disto}}(y) + y^2 \sigma_n^2} \right) \qquad (4.68)$$

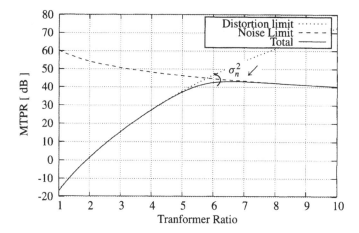

Figure 4.20: Evaluation of the upper limit of the achievable MTPR, taken the noise and distortion influences into account

The results of this calculation are depicted in figure 4.20 for a reasonable output noise level of -120 dBm/Hz. It can be clearly observed that with a zeroth order SOPA the G-Lite specifications (see table 2.5) can be met but full ADSL seems to be unreachable with this architecture, unless the very stringent noise-specifications can be met. Note that the output-currents will increase inversely proportional with the square of the transformer ratio. Hence the supply and substrate noise will also increase since this is a switching power amplifier. In section 4.3 on page 122 an improvement on the zeroth order scheme will be proposed that relaxes this limit.

2.3.3 Signal Bandwidth

The signal bandwidth can be regarded in two fashions :

- What is the maximum frequency for which the approximations of the SOPA dithering no longer holds. This criterion can be regarded as the question : which signals are that fast that the SOPA can no longer be regarded as 'time-invariant'.

- Since the signal modulates the square wave output, inter-modulation terms are observable around the limit cycle frequency. The bandwidth can in this case be regarded as the highest frequency for which the inter-modulation tones within the bandwidth become too large with respect to the noise-floor.

Both considerations will be further elaborated below.

The first, time-invariance consideration is similar to the filter hypothesis. This effect cannot be calculated by regarding the TSIDF, since frequency dependence is omitted by definition. However, if we consider an exponential tail

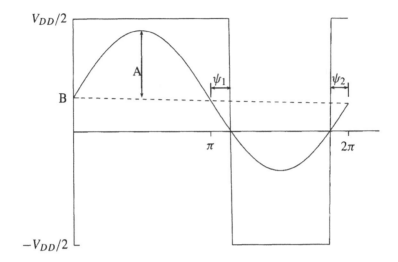

Figure 4.21: Waveforms to calculate dual-input describing function for sinu-soidal plus exponential inputs.

in the input wave-form, a DIDF for a sinusoidal plus exponential can be calcu-lated. Figure 4.21 represents the input and output waveforms used to calculate this describing function. The describing function for an ideal comparator can be calculated from the integral representation [Gelb and Vander Velde, 1968].

$$N_B(A, B) = \frac{1}{B} \int\limits_0^{\frac{1}{f_1}} y \left(B \exp\left(-\frac{t}{\tau}\right) + A \sin\left(2\pi f_1 t\right) \right) dt \qquad (4.69)$$

In which $y(x)$ denotes the input-output relationship of the nonlinear element and f_1 is the limit cycle frequency. For an ideal comparator and following the notations of figure 4.21, this can be elaborated to :

$$N_B(A, B) = \frac{1}{B} \left[\int\limits_0^{\frac{1}{2f_1} + \psi_1} \frac{V_{DD}}{2} dt - \int\limits_{\frac{1}{2f_1} + \psi_1}^{\frac{1}{f_1} - \phi_2} \frac{V_{DD}}{2} dt + \int\limits_{\frac{1}{f_1} - \psi_2}^{\frac{1}{f_1}} \frac{V_{DD}}{2} dt \right] \qquad (4.70)$$

This integral can be easily solved if the crossing points ϕ_1 and ϕ_2 are known. This implies solving the following equation around $t = 1/(2f_1)$ and $t = 1/(f_1)$:

$$B \exp\left(-\frac{t}{\tau}\right) + A \sin\left(2\pi f_1 t\right) \qquad (4.71)$$

The values for ϕ_1 and ϕ_2 can be found by approximating (4.71) by the Taylor series expansions around $1/(2f_1) + \phi_1$ and $1/(f_1) + \phi_2$. This gives the following solutions :

$$\psi_1 = \frac{B\tau \exp\left(-\frac{1}{2f_1\tau}\right)}{2\pi A f_1 \tau + B \exp\left(-\frac{1}{2f_1\tau}\right)} \tag{4.72}$$

$$\psi_2 = \frac{B\tau \exp\left(-\frac{1}{f_1\tau}\right)}{2\pi A f_1 \tau - B \exp\left(-\frac{1}{f_1\tau}\right)} \tag{4.73}$$

Filling in solutions (4.72) and (4.73) in the limits of (4.70) gives :

$$N_B(A, B) = \frac{V_{DD} f_1}{2}\left[\frac{2\tau \exp\left(-\frac{1}{2f_1\tau}\right)}{2\pi A f_1 \tau + B \exp\left(-\frac{1}{2f_1\tau}\right)}\right.$$
$$\left. + \frac{2\tau \exp\left(-\frac{1}{f_1\tau}\right)}{2\pi A f_1 \tau - B \exp\left(-\frac{1}{f_1\tau}\right)}\right] \tag{4.74}$$

Note that for $\tau \to \infty$ this simplifies to the regular TSIDF. To calculate the time-variance induced distortion, the expression of (4.74) is series expanded.

$$N_B(A, B) = \frac{V_{DD}}{2\pi A}\left[\left(\exp\left(-\frac{1}{2f_1\tau}\right) + \exp\left(-\frac{1}{f_1\tau}\right)\right)\right.$$
$$+ \left(\frac{\exp\left(-\frac{2}{f_1\tau}\right) - \exp\left(-\frac{1}{f_1\tau}\right)}{2\pi f_1\tau A}\right) B$$
$$\left. + \left(\frac{\exp\left(-\frac{3}{2f_1\tau}\right) + \exp\left(-\frac{3}{f_1\tau}\right)}{(2\pi f_1\tau A)^2}\right) B^2 + O\left(B^3\right)\right] \tag{4.75}$$

Since $f_1\tau \to \infty$, the exponentials have an upper-limit of 1. Comparing the B^2 term of (4.75) with the corresponding term in (4.53),i.e. comparing the third order distortion generating terms, will give a limit for the minimal exponential time constant τ. If the SOPA is distortion limited (Cf. figure 4.20), and we set the distortion contribution of the non-linearity to be at least as important as the time-variance induced distortion, the following inequality needs to be fulfilled :

$$\frac{V_{DD}}{2\pi A}\frac{2}{(2\pi f_1\tau A)^2} \leq \frac{V_{DD}}{8\pi A^3} \tag{4.76}$$

which can be simplified to :

$$\tau_{min} \geq \frac{2\sqrt{2}}{2\pi f_1} \approx 3\tau_1 \tag{4.77}$$

This calculation corresponds with the rule-of-thumb $\tau_{min} \geq \frac{3}{\omega_1}$ as given by [Gelb and Vander Velde, 1968], on pages 318-328. The limit for the input signals and thus the maximum bandwidth for a distortion limited zeroth order SOPA is thus a third of the limit cycle frequency. Compared with other over-sampling techniques like discrete time $\Delta\Sigma$-modulation [Candy and Temes, 1992], this low Over Switching Ratio (OSR)[2] is one of the major advantages of the SOPA technique.

The second consideration, being the intrusion of the modulation products of the baseband signal into the bandwidth has been described in [Roza, 1997]. For this analysis the output of the SOPA is regarded to be a square wave with modulated duty cycle and modulated temporal frequency. An example of such a modulation scheme is a modulated square wave where the duty cycle (D), being the pulse width (α) divided by the instantaneous period (T) is modulated by the instantaneous signal amplitude ν :

$$D = \frac{\alpha}{T} = \frac{\nu + 1}{2} \tag{4.78}$$

The instantaneous pulsation (ω) is modelled as :

$$\frac{\omega}{\omega_1} = 1 - \left(\frac{\nu}{V_{DD}}\right)^2 \tag{4.79}$$

It is shown in [Roza, 1997] that the modulation of an unclocked feedback loop with a linear filter and a comparator at the output for which the linear filter has a decaying transfer function for frequencies above the mean switching frequency will converge to this modulation. Examples of such systems include any order SOPA amplifier and an asynchronous $\Delta\Sigma$-modulator. Moreover it can also be proven that the modulation described by equations (4.78) and (4.79) becomes a better approximation for higher order frequency roll-off, thus for higher loop filter orders and/or higher SOPA orders. The spectral content for this ideal modulation can be calculated starting from the Fourier series expansion of a static[3] square wave $s(t)$ with a frequency ω_1 and a duty cycle α_1/T_1[4] :

$$s(t) = \left(2\frac{\alpha_1}{T_1} - 1\right) + 4\sum_{k=1}^{\infty} \frac{\sin\left(k\pi\frac{\alpha_1}{T_1}\right)}{k\pi} \cos\left(k\omega_c t + \phi_i\right) \tag{4.80}$$

---

[2]Note that we use the term overswitching ratio and not oversampling ratio since in a SOPA no sampling step is performed. For synchronised systems the two terms are used as synonyms.

[3]It is also proven in the appendices of [Roza, 1997] that the approximations also hold for dynamic input signals.

[4]The subscript 1 is explicitly added to denote that the central frequency in the square wave is the limit cycle frequency.

In a SOPA, the duty cycle of this square wave can be considered to be modulated by (4.78). The modulated frequency (4.79) can be calculated as a time-independent frequency deviation. For this calculation the instantaneous signal amplitude is modelled as :

$$v = B \cos(\omega_{sig} t) \tag{4.81}$$

which leads to the following modulated frequency :

$$\begin{align}
\omega &= \omega_1 \left(1 - \left(\frac{B}{V_{DD}} \right)^2 \cos^2(\omega_{sig} t) \right) \tag{4.82} \\
&= \omega_1 \left(1 - \frac{1}{2} \left(\frac{B}{V_{DD}} \right)^2 \right) + \frac{\omega_1}{2} \left(\frac{B}{V_{DD}} \right)^2 \cos(2\omega_{sig} t) \tag{4.83} \\
&= \omega_c + \omega_\Delta \tag{4.84}
\end{align}$$

The time-independent frequency deviation (ω_Δ) will lead to a phase angle modulation. This phase modulation (ϕ_i) is calculated by integrating the instantaneous frequency deviation [MacLachlan, 1955].

$$\phi_i = \frac{\omega_1}{4\omega_{sig}} \left(\frac{B}{V_{DD}} \right)^2 \sin(2\omega_{sig} t) \tag{4.85}$$

Substituting (4.83) and (4.85) in (4.80) gives :

$$\begin{align}
s_{\mathrm{mod}}(t) &= B \cos \left(\omega_{sig} t \right) \\
&+ 4 \sum_{k=1}^{\infty} \frac{\sin \left(k\pi \left(\frac{B/V_{DD} \cos(\omega_{sig} t) + 1}{2} \right) \right)}{k\pi} \\
&\times \cos \left(k\omega_1 t \left(1 - \frac{1}{2} \left(\frac{B}{V_{DD}} \right)^2 \right) + \frac{\omega_1}{4\omega_{sig}} \left(\frac{B}{V_{DD}} \right)^2 \sin(2\omega_{sig} t) \right)
\end{align}$$

$$\tag{4.86}$$

Using the Werner formulas [Weisstein,], the modulated waveform can be rewritten as:

$$s_{mod}(t) = B\cos\left(\omega_{sig}t\right)$$

$$+2\sum_{k=1}^{\infty} \frac{\sin\left(\frac{k\pi}{2} + \frac{k\pi B}{2V_{DD}}\cos(\omega_{sig}t) + \frac{\omega_1 B^2}{4\omega_{sig}V_{DD}^2}\sin(2\omega_{sig}t)\right)}{k\pi}$$

$$\times\cos\left(\left[1 - \frac{B^2}{2V_{DD}^2}\right]k\omega_1 t\right)$$

$$+2\sum_{k=1}^{\infty} \frac{\sin\left(\frac{k\pi}{2} + \frac{k\pi B}{2V_{DD}}\cos(\omega_{sig}t) - \frac{\omega_1 B^2}{4\omega_{sig}V_{DD}^2}\sin(2\omega_{sig}t)\right)}{k\pi}$$

$$\times\cos\left(\left[1 - \frac{B^2}{2V_{DD}^2}\right]k\omega_1 t\right)$$

$$-2\sum_{k=1}^{\infty} \frac{\cos\left(\frac{k\pi}{2} + \frac{k\pi B}{2V_{DD}}\cos(\omega_{sig}t) + \frac{\omega_1 B^2}{4\omega_{sig}V_{DD}^2}\sin(2\omega_{sig}t)\right)}{k\pi}$$

$$\times\sin\left(\left[1 - \frac{B^2}{2V_{DD}^2}\right]k\omega_1 t\right)$$

$$+2\sum_{k=1}^{\infty} \frac{\cos\left(\frac{k\pi}{2} + \frac{k\pi B}{2V_{DD}}\cos(\omega_{sig}t) - \frac{\omega_1 B^2}{4\omega_{sig}V_{DD}^2}\sin(2\omega_{sig}t)\right)}{k\pi}$$

$$\times\sin\left(\left[1 - \frac{B^2}{2V_{DD}^2}\right]k\omega_1 t\right) \tag{4.87}$$

The following conclusions can be drawn from this equation :

- If equation (4.86) is regarded as a time domain signal, the equations suggest the spectrum looks like a summation of QAM signals. The spectra are up-converted to $\left[1 - B^2/(2V_{DD}^2)\right]\omega_1 t$ and its higher harmonics. The amplitude in a channel is inversely proportional with the channel number k, thus the energy in these modulation channels drops inversely proportional with k^2. This is in accordance with the observations of the numerical simulations of figure 4.3(a).

- Remarkable is that the mean switching frequency shifts with the energy in the driving signal. The mean switching frequency is thus dependent on the limit cycle frequency, but also on the energy content of the driving signal. To derive the maximum bandwidth of the SOPA the rms output power needs

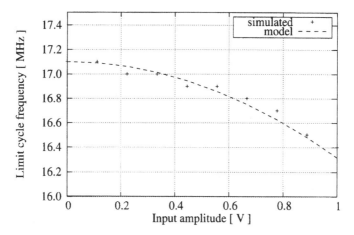

Figure 4.22: Deviation with signal amplitude of the mean switching frequency from the limit cycle frequency. All independent values are taken the defaults as defined in table 4.1

also be taken into account :

$$\omega_c = \left(1 - \frac{B^2}{2V_{DD}^2}\right)\omega_1 \tag{4.88}$$

This effect is plotted and compared with numerical verifications in figure 4.22.

- The resulting square wave can thus be expressed as :

$$s_{\text{mod}}(t) = v + \sum_{k=1}^{\infty} s_{k_c}(t)\cos(k\overline{\omega}t) + \sum_{k=1}^{\infty} s_{k_s}(t)\sin(k\overline{\omega}t) = v + \sum_{k=1}^{\infty} s_k(t) \tag{4.89}$$

Applying the Bessel function series expansions [Abramowitz and Stegun, 1972]

$$\cos(z\sin\theta) = J_0(z) + \sum_{p=1}^{\infty} J_{2p}(z)\cos(2p\theta) \tag{4.90}$$

$$\sin(z\sin\theta) = 2\sum_{p=1}^{\infty} J_{2p-1}(z)\sin((2p-1)\theta) \tag{4.91}$$

to (4.86), leads to the following spectrum for the ith harmonic band [Green and Williams, 1992] :

$$
\begin{aligned}
s_i(t) &= \frac{V_{DD}}{i\pi} \sin\left(i\frac{\pi}{2}\right) \sum_{n=-\infty}^{\infty} \sum_{k=-\infty}^{\text{even}} J_n\left(i\frac{B\pi}{2V_{DD}}\right) J_k\left(i\frac{\omega_1 B^2}{4\omega_{\text{sig}} V_{DD}^2}\right) \\
&\quad \cos\left(\omega_c t + (2n+k)\omega_{\text{sig}} t\right) \\
&= \frac{V_{DD}}{i\pi} \cos\left(i\frac{\pi}{2}\right) \sum_{n=-\infty}^{\infty} \sum_{k=-\infty}^{\text{odd}} J_n\left(i\frac{B\pi}{2V_{DD}}\right) J_k\left(i\frac{\omega_1 B^2}{4\omega_{\text{sig}} V_{DD}^2}\right) \\
&\quad \sin\left(\omega_c t + (2n+k)\omega_{\text{sig}} t\right)
\end{aligned}
\tag{4.92}
$$

Note that due to the $\cos\left(i\frac{\pi}{2}\right)$ term, for an odd harmonic band (i=odd), only frequency peaks spaced at even multiples of the signal frequency ω_{sig} from the centre frequency ω_c can occur. This will become important in the description of the forced coupled SOPA in section 4.2.4. The signal amplitude at a certain frequency in (4.92) is a combination of different combinations of n and k. Equation 4.92 can be rearranged using the definition $l = 2n + k$.

$$
s_i(t) = \frac{V_{DD}}{i\pi} \sum_{l=-\infty}^{\infty} \frac{1 - (-1)^{i+l}}{2} B(i,l) \cos(\omega(i,l) + \phi(i))
\tag{4.93}
$$

where

$$
\omega(i,l) = i\omega_c + l\omega_{\text{sig}}
\tag{4.94}
$$

$$
B(i,l) = \sum_{p=-\infty}^{\infty} J_p\left(i\frac{\omega_1 B^2}{4\omega_{\text{sig}} V_{DD}^2}\right) J_{(l-2p)}\left(i\frac{B\pi}{2V_{DD}}\right)
\tag{4.95}
$$

$$
\phi(i) = i\frac{\pi}{2}
\tag{4.96}
$$

Figure 4.23 shows the evaluation of the model calculated in (4.93). The model calculates the spectrum at discrete values, therefor they are depicted as Dirac impulses (arrows). As can be noticed, the model matches the simulated behaviour with a high accuracy, thus it can be used to determine the SOPA's bandwidth.

In principle the modulation bandwidth around the harmonics of the limit cycle frequency stretch from $-\infty$ to ∞. However, due to the fast decay of the Bessel functions only a limited number of tones should be taken into account. From figure 4.23, it can be derived that the tone at $\omega_1 - 4\omega_{\text{sig}}$ is already less important than the third order harmonic due to the non-linearity of the system. The maximum bandwidth of the SOPA will in this case thus be limited to one sixth of the limit cycle frequency, incorporating the limit cycle shift dictated by (4.88).

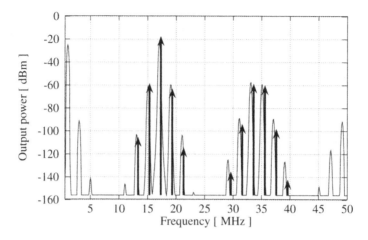

Figure 4.23: Comparison between the calculated sideband power (arrows) and the simulated output spectrum (solid line).

2.4 Driving the coupled system

2.4.1 First observations

The complete zeroth order SOPA system consists of two SOPA building blocks. In this section, a mathematical model for the behaviour of the coupled system will be derived. The model will start from the knowledge gathered from the calculations for the un-coupled system, performed in previous sections and from the calculations of the limit cycling behaviour of the coupled system, done in section 4.2.2.

Figure 4.24 shows four spectra taken at the output node of one SOPA building block in grey and the spectrum taken at the line driver input (i.e. after combining the two outputs with a signal transformer) in black. These spectra result from numerical simulations. Four different coupling factors α where taken, representing distinct coupling regions : $\alpha = 0$ for the uncoupled system, $\alpha = 0.05$ for low coupling and active output synthesis, $\alpha = 0.25$ for resistive output termination and $\alpha = 0.35$ for strong coupling. The following observations are clear from first inspection of these spectra:

- The output amplitude is doubled due to the bridge configuration. This is beneficial for creating line drivers in low voltage technologies.

- Since the limit cycle oscillation is common mode, it is not transferred towards the load. Due to the higher excessive common mode gain (cf. section 4.2.2 on page 85) for higher coupling factors, the suppression will be better for increasing coupling factors.

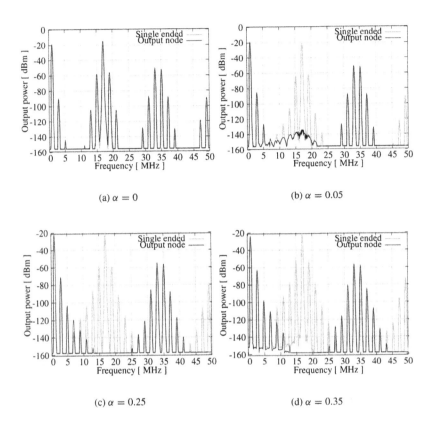

(a) $\alpha = 0$

(b) $\alpha = 0.05$

(c) $\alpha = 0.25$

(d) $\alpha = 0.35$

Figure 4.24: Numerical simulations of the coupled SOPA system for different coupling factors α. Spectra are calculated at the output of one SOPA building block (grey) and at the line input (black)

- Not only the limit cycle frequency is suppressed, also the modulation sidebands in the odd harmonics of the limit cycle frequency. This can be predicted from (4.92), since the modulation peaks on odd harmonics of the limit cycle frequency only consist of even multiples of the input frequency. Since the input signal is differential mode, the even multiples will be common mode. Due to this effect, the bandwidth limitation by the modulation spurs is liquidated.

- By combining the two outputs, the distortion increases. In the next subsection a mathematical model for this decay will be given. This effect grows in importance with increasing coupling factors.

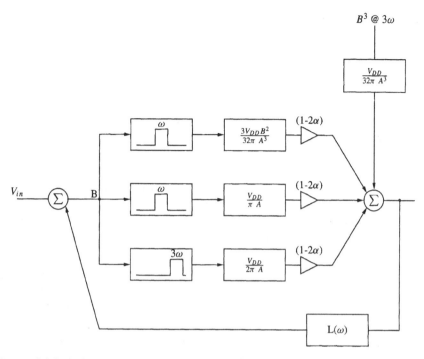

Figure 4.25: Adaption of the model of figure 4.17 to represent the coupling behaviour.

- The modulation bands at the even harmonics are not suppressed, on the contrary next to the amplitude doubling due to the bridge configuration, the sideband energy increases with increasing coupling factors. This growth can be explained by a decreasing loop gain as will be explained in more detail in the next chapter for signal distortion. For the sideband power, this will show up in the $i\frac{B\pi}{2V_{DD}}$ term of (4.92), which is altered to $i\frac{B\pi}{2V_{DD}(1-2\alpha)}$. However, since they reside at higher frequencies and are intrinsically lower than the distortion level, they do not form a new limitation.

- The oscillator pulling towards synchronisation thus offers extra filtering at the cost of an increase in distortion levels.

- Lower coupling factors suffice to suppress the limit cycle and its modulation sidebands without suffering too much decay in linearity. This implies the use of active line termination of putting the resistive termination at the secondary coil of the line transformer.

2.4.2   Model for the forced, coupled system

To calculate the higher distortion due to the coupling, the block schematic of figure 4.17 needs to be adapted to take the coupling into account. The calculation of the resulting distortion takes three distinct steps.

1 The limit cycle frequency and its sidebands are all common mode signals. They can thus be calculated with the un-coupled model [Lindgren, 1964]. The results from the previous sections are thus applicable in the calculation of the dithering self-oscillation.

2 Since an ideal comparator is an odd non-linearity, only odd harmonics will occur. Thus, if the input is a differential, counter mode signal, all generated in-band frequencies will be counter-mode. For this reason, the multivariable system can be simplified to a single input system [Lindgren, 1964] by adding an extra amplification factor $(1 - 2\alpha)$ in the loop. Figure 4.25 shows the resulting block schematic. The next step of the model calculation then consists of calculating the error signal amplitude B at the input frequency ω. This is done by calculating the loop transfer function when only the branches with the bandpass filters at ω are included. The loop-gain consists of two contributions : the linear gain $V_{DD}/(\pi\ A)$ and the contribution $3V_{DD}B^2/(32\pi\ A^3)$ from the down-converted third order distortion. This gain is lowered with an amplification factor $(1 - 2\alpha)$ due to the compensation by the counter phase signal from the other SOPA-building block. The in-band signals are counter phase since the complete SOPA is differentially driven. To derive an expression for the error-signal the real root of the loop gain in function of B needs to be calculated. This cubic equation can be analytically solved by applying Vièta's substitution [Weisstein,].

3 The distortion then is modelled as an error-signal added to the system just behind the comparator. This signal gets suppressed by the loop gain at 3ω. Since the third order distortion signal is in-phase with the signal, the generated distortion signals of the two SOPA-building blocks will also be in counter-phase. The loop gain needs thus also be decreased with a factor $(1 - 2\alpha)$ to take this effect into account.

From this algorithm a closed expression for the distortion in function of the design parameters can be derived. Figure 4.26 shows the resulting model evaluated for values of the coupling factor α from 0 up to 0.5. Simulated values are added with crosses. The model shows a very good fit with the numerical verification. An important conclusion is that there is an important performance decay for higher coupling. This is due to the decreasing loop gain. The implication of this is the need for active impedance synthesis or impedance matching at the secondary of the coupling transformer.

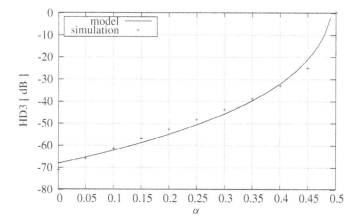

Figure 4.26: Comparison between numerical simulated distortion values and the predicted model for different values of α

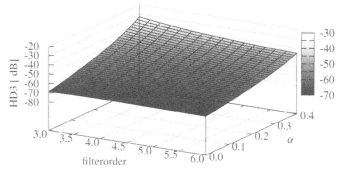

Figure 4.27: Evaluation of the coupled SOPA distortion model for different filter orders and coupling factors

The big advantage of this modelling effort is the fast evaluation of the derived closed formulas, allowing a very fast design space exploration. The distortion model is evaluated for different coupling factors α and different filter orders[5] and is shown in figure 4.27. It is shown that to lower the effect of linearity degradation by direct coupling, the filter order needs to be as high as possible. Since the limit cycle amplitude and frequency is a fixed point in the complex plane by the Barkhausen criterion, a steeper filter characteristic will provide a higher in-band loop gain.

---

[5]Although the filter order is in stricto sensu an integer number, it can be evaluated as a real number for equal positioning of the filter poles. In this case, it should be regarded as a specific frequency offset/amplitude roll-off value at the limit cycle frequency

2.5 Inherent adaptivity of limit cycle systems

Equations (4.53) and (4.57) show the dithering effect of the limit cycle oscillation on the SOPA's inherent non-linearity. This effect can be expanded to other non-idealities of the system. As an example, the finite gain of the comparator is covered in this section.

For this the ideal comparator model, used in the previous sections is replaced with :

$$V_{out} = \begin{cases} \text{GAIN} \cdot x & \text{if } -\delta < x < \delta \\ V_{DD} & \text{if } |x| > \delta \end{cases} \tag{4.97}$$

Note that due to this definition, δ, V_{DD} and the GAIN are coupled by $2\delta \cdot \text{GAIN} = V_{DD}$. The describing function for this comparator is given by:

$$N(A) = \text{GAIN} \frac{2}{\pi} \left[\arcsin\left(\frac{\delta}{A}\right) + \frac{\delta}{A}\sqrt{1 - \left(\frac{\delta}{A}\right)^2} \right] \tag{4.98}$$

This can be approximated as :

$$N(A) = \frac{2V_{DD}}{\pi A} - \frac{V_{DD}\delta^2}{3\pi A^3} + O\left(\delta^4\right) \tag{4.99}$$

Filling this expression in the Barkhausen criterion, gives a third order equation in the limit cycles amplitude A, which can be easily solved to :

$$A = \frac{2V_{DD}\cos^n\left(\frac{\pi}{n}\right)}{\pi} - \frac{\pi\delta^2}{12V_{DD}\cos^n\left(\frac{\pi}{n}\right)} \tag{4.100}$$

The TSIDF that will be used to calculate the distortion in the same way as it has been explained in figure 4.17. The TSIDF is the result of the following improper integral :

$$N_B(A, B) = \frac{V_{DD}}{\pi\delta B} \int_{-\infty}^{\infty} \frac{\sin(\delta u)}{u^2} J_0(Au) J_1(Bu) \, du \tag{4.101}$$

This can be rewritten by substituting $u = v/A$ to :

$$N_B(A, B) = \frac{V_{DD} A}{\pi\delta B} \int_{-\infty}^{\infty} \frac{\sin\left(v\frac{\delta}{A}\right)}{v^2} J_0(v) J_1\left(v\frac{B}{A}\right) \, dv \tag{4.102}$$

To simplify this expression in a form from which the third order distortion can be derived, the sine-function is replaced by its MacLaurin-expansion. This can be done since δ should always be much smaller than the limit cycle amplitude

A. If not, the comparator will act like a regular linear amplifier and the limit cycle will vanish, since in linear systems limit cycles do not occur.

$$N_B(A, B) = \frac{2V_{DD}\,A}{\pi\,\delta\,B}\left(\int_0^{\infty}\frac{\delta}{A\,v}J_0(v)\,J_1\left(v\frac{B}{A}\right)dv\right.$$

$$\left.-\int_0^{\infty}\frac{\delta^3\,v}{6A^3}J_0(v)\,J_1\left(v\frac{B}{A}\right)dv\right) \quad (4.103)$$

$$= N_B(A, B)|_{\delta=0}$$

$$-\frac{2V_{DD}\,A}{\pi\,\delta\,B}\int_0^{\infty}\frac{\delta^3\,v}{6A^3}J_0(v)\,J_1\left(v\frac{B}{A}\right)dv \quad (4.104)$$

in which $N_B(A, B)|_{\delta=0}$ denotes the distortion terms of the ideal comparator as shown in (4.53). The improper integral is from the Weber-Schafheitlin type [Luke, 1962] and can thus be calculated as follows:

$$\int_0^{\infty}t^{\lambda}J_{\mu}(at)\,J_{\nu}(bt)dt =$$

$$\begin{cases} \dfrac{(b/a)^{\nu}(a/2)^{\lambda-1}\Gamma\left(\frac{\mu+\nu-\lambda+1}{2}\right)}{2\Gamma(\nu+1)\Gamma\left(\frac{\mu-\nu+\lambda+1}{2}\right)}\,{}_2F_1\left(\frac{\mu+\nu-\lambda+1}{2},\ \frac{\nu-\mu-\lambda+1}{2};\nu+1;\left(\frac{b}{a}\right)^2\right) \\ \qquad\qquad\text{for}:0<b<a \\[2ex] \dfrac{(a/b)^{\mu}(b/2)^{\lambda-1}\Gamma\left(\frac{\mu+\nu-\lambda+1}{2}\right)}{2\Gamma(\mu+1)\Gamma\left(\frac{\nu-\mu+\lambda+1}{2}\right)}\,{}_2F_1\left(\frac{\mu+\nu-\lambda+1}{2},\ \frac{\mu-\nu-\lambda+1}{2};\mu+1;\left(\frac{a}{b}\right)^2\right) \\ \qquad\qquad\text{for}:0<a<b \end{cases}$$

$$(4.105)$$

With the series development of the hyper-geometric series function (4.51), the TSIDF can be rewritten as :

$$N_B(A, B) = \frac{V_{DD}\,A}{\pi\,\delta\,B}\left[\left(\frac{\delta}{A}-\frac{\delta^3}{12A^3}\right)\frac{B}{A}+\left(\frac{\delta}{8}-\frac{\delta^3}{48A^3}\right)\frac{B^3}{A^3}\right] \quad (4.106)$$

This equation can be used to alter the terms in figure 4.17, from which the third harmonic distortion can be calculated in function of decreasing comparator gain, or alternatively, the increasing saturation point δ. The result of this calculation is shown in figure 4.28, together with numerical simulations of the same system. The obtained model is within .25 dB of the numerical simulations. This is partially due to the anti-aliasing filter that is used in the post-processing of the results of the numerical simulator. The trend-lines predicted

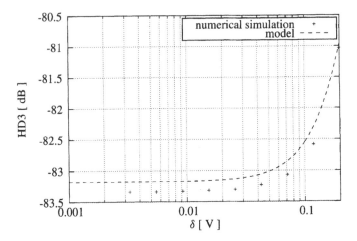

Figure 4.28: Third harmonic distortion of a zeroth order SOPA with third order loop filter versus $\delta = \frac{V_{DD}}{\text{GAIN}}$, for an input amplitude of 0.1 and input frequency of $1/(3\sqrt{3})$ times the limit cycle frequency

by the model are perfectly matched by the numerical results. The significant indifference of the third (and most dominant) harmonic distortion component on the comparator gain is quite surprising. The increase is limited to 3 dB for ranges of the saturation point up to the limit cycle amplitude at infinite gain, thus up to the point were the limit cycle oscillation disappears. This can only be explained by the adaptivity of the system due to the dithering signal.

This can be shown by introducing an extra term ($L(\omega, \zeta)$) to the open loop transfer function which models the impairment (ζ) like non-ideal comparators or line influences on the system. The open loop transfer function (OLTF(ω)) thus can be rewritten as:

$$\text{OLTF}(\omega) = N_B(A, B)H(omega)L(\omega, \zeta) \qquad (4.107)$$

Note that for small values of the error-signal with amplitude B, the TSIDF can be approximated as:

$$N_B(A, B) \approx \frac{N_A(A)}{2} \qquad (4.108)$$

The combination of (4.108) and (4.107) shows that the open loop transfer function for a forced signal always passes through the fixed point $1/2\angle\pi$ since the limit cycle oscillation with amplitude A_0 and frequency ω_0 fulfils the Barkhausen criterion. The open loop transfer function can thus be rewritten as:

$$\text{OLTF}(\omega) = \frac{H(\omega)L(\omega, \zeta)}{2H(\omega_0)L(\omega_0, \zeta)} \qquad (4.109)$$

If the impairment has only a weak dependency on the frequency, meaning $L(\omega, \zeta) \approx L(\zeta)$, the influence of the impairments completely vanishes.

This phenomenon is called 'inherent adaptivity' and can be physically explained as follows. Since for the SOPA amplifier to be in its linear region, the signal levels are chosen in such a way that the comparator is only dependent on the limit cycle frequency A. Thus the comparator will be the actuator in the system. If we consider it to be driven by the impairments, the only thing needed for adaptivity is a way to measure these impairments. The dithering limit cycle measures the non-ideality since its amplitude and/or frequency is changed (Cf. (4.100)). In this way the feedback loop is closed for an impairment input. Its influence is heavily suppressed.

The calculation of the signal distortion due to finite comparator gain acts as an illustration of this principle in this section, but the theory is also valid for other comparator non-idealities. A general rule of thumb for the design of a zeroth order SOPA with sufficient inherent adaptivity is to design the SOPA in such a way that the influence of the non-ideality on the limit cycle amplitude is significantly lower than the ideal limit cycle amplitude. Since the calculation of this influence is far more easy than the influence on signal integrity, this can be done very fast. Due to the adaptivity of the system, this guarantees near ideal system behaviour.

A numerical example. will illustrate this property. A standard zeroth order is take with the default values from table 4.1. If an error of 10% is allowed on the limit cycle amplitude of 0.26 V, the minimal comparator gain can be calculated as follows:

$$\Delta A = 0.026 = \frac{\pi \delta^2}{12 \; 3.3 \; cos^n \left(\frac{\pi}{3}\right)} \tag{4.110}$$

$$\Rightarrow \delta = \sqrt{\frac{1.0296 \; cos^n \left(\frac{\pi}{3}\right)}{\pi}} \tag{4.111}$$

giving δ to be equal to 0.2. This is equivalent with a gain of 10. From the calculations presented in Figure 4.28, this would mean a degradation of only 2.2 dB.

2.6   Conclusions on the analysis of the zeroth order SOPA

A zeroth order SOPA is constructed by feeding back the output of a comparator through a loop filter of order n. Since the SOPA principle counts on the dithering effect of a limit cycle oscillation, the conditions for which limit cycles occur and expressions for its frequency and amplitude have been calculated. It is proven that the filter order should be at least three.

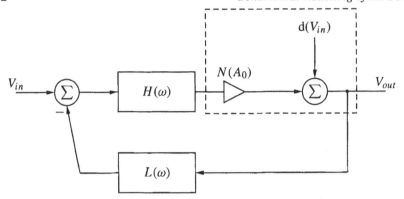

Figure 4.29: Simplified model of a SOPA, using a linearised model for the comparator (dashed box).

Another important feature of the SOPA amplifier is the use of oscillator pulling when two SOPAs are coupled via the load. Due to this effect the limit cycle will not be transferred to the load, if the oscillation are in phase, saving power and extra filtering. Important modelling has been done to get design limits for which the SOPAs are guaranteed to oscillate in phase. Since the line impedance can vary over a wide range in the complex plane, certainly at the relatively high limit cycle frequencies. The result of these calculations for the extreme line impedances, capacitive and inductive loading, has put a maximum of 6 on the order of the loop filter.

In a later section the response on an input signal has been calculated. The model showed a fundamental limit on the achievable MTPR due to up transformation of the noise. As an illustration of the inherent adaptivity property of the SOPA amplifier, this calculation has been repeated for a comparator with finite gain. It has been shown that if the limit cycle oscillation is little disturbed by the impairment, the SOPA will act as if the impairment is not present.

3. Higher order SOPA amplifiers

3.1 Noise shaping technique

As it has been shown in section 4.2.3 on page 98, the input signal amplitude of a zeroth order SOPA needs to be small in order to obtain sufficiently low distortion levels. The consequence of this together with the specified output levels is a high transformer ratio to be used, which severely increases the noise specifications of the amplifier (see figure 4.20). In order to relax these specifications, a noise shaping technique, not unlike the one used in $\Delta\Sigma$-converters [Candy and Temes, 1992] is introduced in the SOPA principle. Figure 4.29 shows a simplified model that will be used to explain the noise shaping technique. If we calculate the transfer from the input to the output, we will find

:

$$\frac{V_{out}}{V_{in}} = \frac{H(\omega)N(A_0)/2}{H(\omega)L(\omega)N(A_0)/2 + 1} \qquad (4.112)$$

In (4.112) A_0 denotes the limit cycle amplitude. The assumption of (4.108) is assumed to also hold for this analysis. Since the limit cycle oscillation fulfils the Barkhausen criterion and $H(\omega)$ has a higher gain for lower frequencies, the term $H(\omega)L(\omega)N(A_0)/2$ will always be larger than $1/2$ for frequencies lower than the limit cycle frequency. The input output relation thus will approximately be a unit gain relationship. The same calculation for the transfer of the distortion term $(d(V_{in}))$ to the output gives:

$$\frac{V_{out}}{d(V_{in})} = \frac{1}{H(\omega)L(\omega)N(A_0)/2 + 1} \qquad (4.113)$$

This equation will tend to zero for high loop gain of the forward filter $H(\omega)$ and frequencies below the limit cycle frequency. Since an ADSL contains a large number of tones which are uncorrelated, we can approximate the contribution of distortion to the ADSL-signal as a white noise source σ_d. The output can thus be described as :

$$
\begin{aligned}
V_{out} &= \frac{H(\omega)N(A_0)/2}{H(\omega)L(\omega)N(A_0)/2 + 1}V_{in} + \frac{1}{H(\omega)L(\omega)N(A_0)/2 + 1}\sigma_d \\
&= \frac{H(\omega)}{H(\omega)L(\omega) + 2\,H(\omega_0)L(\omega_0)}V_{in} \\
&\quad + \frac{2\,H(\omega_0)L(\omega_0)}{H(\omega)L(\omega) + 2\,H(\omega_0)L(\omega_0)}\sigma_d \qquad (4.114)
\end{aligned}
$$

Since the loop filter is low-pass, $L(\omega)$ can be approximated to 1. By stating $H(\omega)L(\omega) \gg 2\,H(\omega_0)L(\omega_0)$, (4.113) is approximated to :

$$V_{out} \approx V_{in} + \frac{H(\omega_0)}{H(\omega)}\sigma_d \qquad (4.115)$$

The output distortion is thus shaped by the inverse of the forward filters. An integrator is thus an ideal forward filter for its high in-band gain. In the same way, noise-shaping is constructed in a $\Delta\Sigma$-converter.

In the next of this section, the calculations of the zeroth order SOPA will be redone to fully investigate the effects of the introduction of forward integrators.

3.2   Limit cycle oscillation

A first observation can be directly made by drawing the modified Nyquist plots as introduced in section 4.2.1.2. The introduction of an integrator in the loop is represented by the rotation of the zeroth order characteristic as the ones of figure 4.6 by $\pi/2$ for every integrator. This is shown in figure 4.30. It is

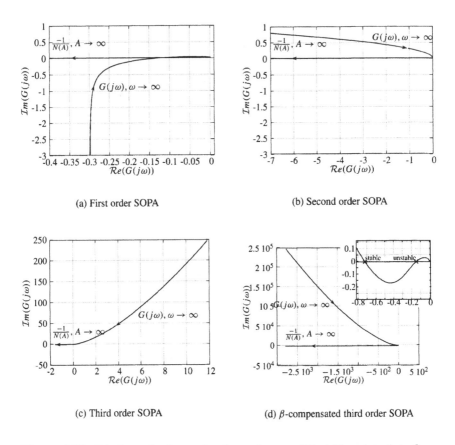

(a) First order SOPA

(b) Second order SOPA

(c) Third order SOPA

(d) β-compensated third order SOPA

Figure 4.30: Limit cycle determination using modified Nyquist plots for uncompensated first (a), second (a) and third order SOPA (c), next to a compensated third order SOPA (d) zoomed in into the crossing region

clear from figure 4.30 that there does not exist a solution for a SOPA which order equals 2 for a third order loop filter. Actually, there exists a solution at infinity, but this is physically impossible. With the order of a SOPA being represented by n, the limit cycle frequency can be calculated as follows :

$$\omega_1|_{\text{no feedback}} = \omega_{fil} \tan\left(\frac{(2-n)\pi}{2m}\right) \qquad (4.116)$$

This equations gives the values of n and m for which no solutions can be found, meaning no limit cycle can be obtained. This limitation can be relaxed by setting the β_i terms nonzero. A closer inspection of figure 4.2 then reveals the different functionality of the feedback terms β_i

β_1, **being the first feedback branch,** sets the output signal level. It is the main feedback branch and its linearity and noise-level will determine the complete systems behaviour.

β_n, **the feedback branch closest to the comparator,** introduces mathematical zeros at f_{int}/β_n, with f_{int} the unit gain of the integrators. This is due to the fact that for low loop gains the feedback branch of β_n short circuits the other integrators. In this way the final β_n loop will determine the limit cycle oscillation for low f_{int}-values.

the other β_i terms are introduced to set the signal levels at the outputs of the different integrators. Clipping of an integrator should be certainly avoided since the insertion of another non-linearity in the loop will destroy the dithering effect of the limit cycle oscillation.

The result of this β-compensation on the modified Nyquist plot can be seen in figure 4.30(d). The introduction of the extra zeros introduces two solutions to the Barkhausen criterion, where the un-compensated modified Nyquist plot (see figure 4.30(c)) does not exhibit a solution . The Nyquist criterion states that the first will be the stable one. Going to higher integrator bandwidths, the obtained curve rotates clockwise around the origin. As a consequence, the solutions shift towards higher limit cycle amplitudes until the $G(\omega)$ curve is tangent to the $N(A)$ curve. Above this integrators Gain Bandwidth (GBW), the SOPA is always unstable.

To calculate the limit cycles frequency and amplitude, the following route is followed. Firstly, it is assumed that the unit gains of the integrators are low enough for the β_n zeros to have full effect. As a result the SOPA is assumed to be of first order for what the limit cycle is concerned. The resulting limit cycle frequency can than be easily calculated, since the integrator introduces a complete $\pi/2$ phase shift.

$$\omega_1|_{n=1} = \omega_{fil} \tan\left(\frac{\pi}{2m}\right) \tag{4.117}$$

To include the effect of the introduced β_n zeros, the complete phase equation is written out.

$$Ph(\omega_1) = n \arctan\left(\frac{\omega_1}{\omega_{fil}}\right) + \frac{m\,\pi}{2} - (m-1) \arctan\left(\frac{\omega_1}{\omega_{int}}\right) - \pi \tag{4.118}$$

To calculate a model for the limit cycle frequency, the first order Taylor series approximation of this function is calculated for frequencies around the first order SOPA limit cycle frequency given in (4.117). The limit cycle frequency is thus :

$$\omega_1 = \omega_1|_{n=1} - \frac{Ph(\omega_1|_{n=1})}{Ph'(\omega_1|_{n=1})} \tag{4.119}$$

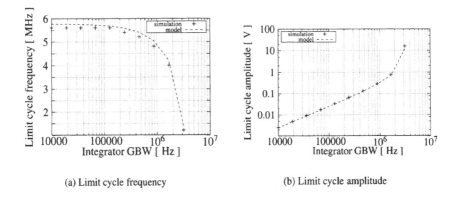

(a) Limit cycle frequency   (b) Limit cycle amplitude

Figure 4.31: Evaluations of the limit cycle frequency and amplitude for a second order sopa versus integrator GBW

Although this equation becomes too large to properly be printed due to the increased number of independent parameters, it is easily evaluated. This expression, like the others which follow below, are explicit, non-recursive equation which give a one-to-one relationship between all design parameters as defined in figure 4.2 and the calculated performance parameter. Evaluation is thus always very fast. The results are shown in figure 4.31(a) and compared with numerical simulations. The limit cycle frequency can then be easily solved by filling in ω_1 in the loop gain.

$$A_1 = \frac{2V_{DD}}{\pi} \left(\frac{\omega_{fil}}{\sqrt{\omega_{fil}^2 + \omega_1^2}} \right)^n \left(\frac{\sqrt{\left(\frac{\omega_{int}}{\beta_n}\right)^2 + \omega_1^2}}{\frac{\omega_{int}}{\beta_n}} \right)^{(m-1)} \left(\frac{\omega_{int}}{\omega_1} \right)^m \quad (4.120)$$

This formula is evaluated in figure 4.31(b) and compared with numerical simulations. From this figure a first limit for the integrators GBW can be derived. If the integrators unit gain frequency is too close to the limit cycle frequency, the limit cycle amplitude will saturate the integrators output, introducing a second non-linearity in the loop. In the curves of figure 4.31 clearly two distinct regions can be observed.

A first region, roughly below 1 MHz integrator GBW in this example, is the zone where the zeros have settled completely. The limit cycle frequency can be calculated by (4.117). The limit cycle frequency exhibits the first order decay of the integrator right in front of the comparator. Above this limit the phase shift of the previous stages is still not completely compensated by the short circuit of β_n. The limit cycle frequency drops with the inverse tangent of the frequency spacing and the limit cycle amplitude increases almost exponentially

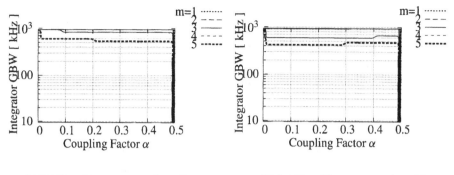

(a) Stability of the common mode oscillation

(b) Stability of the counter mode oscillation

Figure 4.32: Stability region boundaries for different SOPA-orders m in the integrator GBW / coupling factor plane

since the loop gain increases with decreasing limit cycle frequency and with higher integrator GBW. It is thus advisable to construct the SOPA in such a way that the influence of the first integrators towards the limit cycle amplitude is minimised, since the amplitude change due to mismatch in the GBW of the integrators will quickly lead the output of the final integrator stage towards saturation.

3.3 Coupled System Equations

3.3.1 Resistive Coupling

It is shown that for a zeroth order SOPA under resistive coupling the common mode oscillation is dominant, i.e. the two coupled SOPAs will always oscillate in phase. Also for a higher order SOPA two oscillatory modes are possible : the common mode oscillation with amplitude and frequency independent of α and given in (4.119) and (4.120), and the counter mode oscillation at the same oscillation frequency but with a lower amplitude $A_2 = A_1(1-2\alpha)$.

To calculate the dominant oscillation mode, the stability of the two modes needs to be calculated. Therefor the stability criterion in angular form is calculated for the two modes.

$$\left.\frac{\partial \phi}{\partial \omega}\right|_{A_i,\omega_i} \left.\frac{\partial M}{\partial A}\right|_{A_i,\omega_i} - \left.\frac{\partial \phi}{\partial A}\right|_{A_i,\omega_i} \left.\frac{\partial M}{\partial \omega}\right|_{A_i,\omega_i} > 0 \qquad (4.121)$$

Figure 4.32 shows the boundaries for the stability condition (4.121) for different SOPA-orders. The stability is evaluated in the integrator GBW / coupling factor plane. The stability of the common mode case is connected with the exponential growth of the limit cycle amplitude, as it has been pointed out in figure 4.31(b). In other words: as long as the limit cycle amplitude is bounded compared with the supply voltage, the common mode limit cycle oscillation will be stable. Another important conclusion to be drawn from figure 4.32 is that there exist zones in the integrator GBW / coupling factor plane where the common mode oscillation is the only stable mode. This zone is located in the practical coupling factor half-plane ($\alpha \leq 0.5$), so it can be assumed that due to its higher loop gain, higher order SOPAs do have better coupling properties.

To determine the dominant mode in the region where both oscillatory modes are stable, the ratio between the loop gain for a common mode signal while the system is in counter mode oscillation (TFCommon) over the loop gain for a counter mode signal while the system is in common mode oscillation (TFCounter) needs to be calculated.

$$\text{TFCommon} = \left(\frac{\alpha \left(\frac{\omega_c}{I\omega_1 + \omega_c} \right)^n \left(\frac{\omega_u}{I\omega_1} \right)^m \left(\frac{I\omega_1 + \omega_z}{\omega_z} \right)^{(m-1)} \frac{V_{DD}}{\pi\, A_2}}{1 + (1 - \alpha) \left(\frac{\omega_c}{I\omega_1 + \omega_c} \right)^n \left(\frac{\omega_u}{I\omega_1} \right)^m \left(\frac{I\omega_1 + \omega_z}{\omega_z} \right)^{(m-1)} \frac{V_{DD}}{\pi\, A_2}} \right)^2$$

(4.122)

Taking into account that (ω_2, A_2) is a solution of the Barkhausen criterion and $\omega_2 = \omega_1$, the following holds :

$$\left(\frac{\omega_c}{I\omega_1 + \omega_c} \right)^n \left(\frac{\omega_u}{I\omega_1} \right)^m \left(\frac{I\omega_1 + \omega_z}{\omega_z} \right)^{(m-1)} \frac{V_{DD}}{\pi\, A_2} = \frac{-1}{1 - 2\alpha}$$

(4.123)

which simplifies TFCommon to :

$$\text{TFCommon} = \left(\frac{\alpha}{1 - 3\alpha} \right)^2$$

(4.124)

The same calculations can be done with TFCounter :

$$\text{TFCounter} = \left(\frac{\alpha \left(\frac{\omega_c}{I\omega_2 + \omega_c} \right)^n \left(\frac{\omega_u}{I\omega_2} \right)^m \left(\frac{I\omega_2 + \omega_z}{\omega_z} \right)^{(m-1)} \frac{V_{DD}}{\pi\, A_1}}{1 + (1 - \alpha) \left(\frac{\omega_c}{I\omega_2 + \omega_c} \right)^n \left(\frac{\omega_u}{I\omega_2} \right)^m \left(\frac{I\omega_2 + \omega_z}{\omega_z} \right)^{(m-1)} \frac{V_{DD}}{\pi\, A_1}} \right)^2$$

(4.125)

Taking into account that (ω_1, A_1) is a solution of the Barkhausen criterion for the non-coupled case, means:

$$\left(\frac{\omega_c}{I\omega_1 + \omega_c} \right)^n \left(\frac{\omega_u}{I\omega_1} \right)^m \left(\frac{I\omega_1 + \omega_z}{\omega_z} \right)^{(m-1)} \frac{V_{DD}}{\pi\, A_1} = -1$$

(4.126)

which simplifies TFCounter to :

$$\text{TFCommon} = \left(\frac{\alpha}{1 + \alpha} \right)^2$$

(4.127)

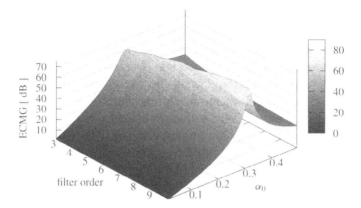

Figure 4.33: The ECMG for a differential SOPA driving an inductive line

Dividing the two will lead to :

$$\frac{\text{TFCommon}}{\text{TFCounter}} = \left(\frac{1 + \alpha}{1 - 3\alpha}\right)^2 \tag{4.128}$$

which is always larger than 1 for practical values of the coupling factor α. This means that for resistive loads, the common oscillatory mode is always the dominant one. Note also that this expression for the ECMG is the same as for the zeroth order SOPA. Due to the Barkhausen criterion and the connection between the TSIDF and the single input DF for the normal working region of the SOPA, this was to be expected.

3.3.2 Non-Resistive Coupling

The same calculations can be performed for the non-resistive line models as it has been done for the zeroth order SOPAs in section 4.2.2.2.

The inductive line model pushes the counter mode oscillation towards higher frequencies. Therefor the common mode gain will be higher than the counter mode one. Figure 4.33 shows the calculated ECMG for a first order SOPA amplifier. The ECMG is almost the square of the zeroth order ECMG, especially for the region in which impedance matching is important (higher α_0 values). The role of the filter order becomes negligible.

For the capacitive line model, the DC coupling factor α_0 is of an utmost importance when taking the $(1 - 2\alpha)$ filter characteristic of figure 4.15 into consideration. For high coupling factors, the extra filter is very low frequent and accounts for a $\pi/2$ phase shift which renders the counter mode oscillation unstable. For low DC coupling factors however the $(1 - 2\alpha)$

filter renders a low frequent counter mode self oscillation. Due to the $1/\omega$ characteristic of the SOPA loop, the counter mode oscillation will become dominant. The counter mode limit cycle amplitude however will enter the exponentially increasing zone as depicted in figure 4.31(b). Therefor the integrator will saturate in a real design. This will lead to a limit cycle oscillation which is suppressed towards the line but not completely. The two SOPAs oscillate with a small phase shift between them. If this is unwanted, the designer can either add an extra zero in the loop or add an extra capacitance at the outputs to pull the corner frequency of the $(1 - 2\alpha)$ sufficiently low, so the counter mode oscillation becomes unstable. Another strategy would be increasing the SOPA-order higher than 1 and placing the integrator GBWs in such a way that the maximum attainable integrator output swing is reached. Since for a higher counter mode oscillation amplitude the gain will drop, only the common mode oscillation will occur. This is similar to the positioning of the integrator GBW in the zone where only the common mode is stable for the resistive case (see figure 4.32).

3.4   Forced System Oscillation

3.4.1   Dynamic Range Calculation

The transition to higher order SOPA amplifiers was inspired by the noise-shaping technique that is used in $\Delta\Sigma$-converters. In this section this effect is put into the dynamic range model of section 4.2.3 to get a quantitative measure for this improvement.

In figure 4.34 the adapted block scheme is presented to calculate the distortion behaviour of a higher order SOPA-amplifier. An elaborated description of the construction of this block scheme has been given in section 4.2.3.2. The difference is the introduction of an extra gain block $G(\omega)$ between the summing node and the comparator input. The results of this calculation for an input signal of 0.1 V, using a third order SOPA is shown in figure 4.35. The noise-shaping behaviour can be clearly observed.

The obtained model is compared with numerical simulations. The model matches almost perfectly the numerical simulations. At higher frequencies, however, the distortion roll-off of the simulations does not seem to keep up with the predicted results, but this is due to the influence of the sideband modulation terms who enter the third harmonic band. At the lowest frequencies, the simulation results are limited by the numerical resolution.

The obtained curve can be divided in three distinct regions :

1 The first region shows the noise-shaping characteristic like it has been predicted in section 4.3.1. This is the favourable operating region. The noise shaping can be used to comply with different standards at once, without consuming too much power, nor chip area. For systems where the bit load-

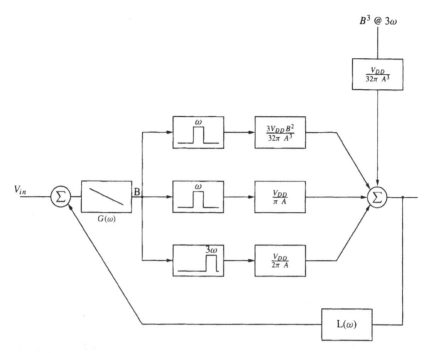

Figure 4.34: Block schematic of a higher order SOPA used to calculate the distortion.

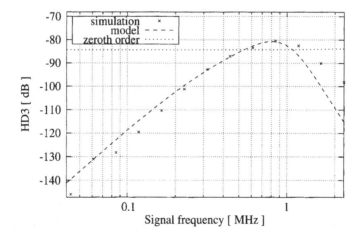

Figure 4.35: Third order harmonic distortion versus input signal frequency for a third order SOPA amplifier. Other parameters have default values (see table 4.1)

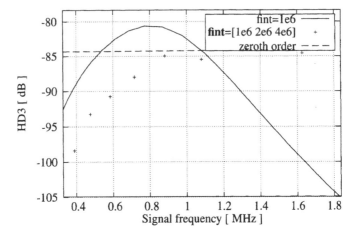

Figure 4.36: Effect of up-scaled integrator GBWs

ing of the higher frequent tones is lower than for the lower frequent tones, like it is the case for xDSL, this solution enables a low voltage, high performance power amplifier that does not degrade the overall bit-error rate.

2 The distortion goes up with higher frequencies, but reaches a maximum before the integrators GBW. In the example of figure 4.35 this maximum is even higher than the obtained distortion for a zeroth order SOPA. This is due to the fact that all integrator unit gain frequencies have the same value. This is not a realistic design point, since the integrators have a different role in the complete system. This can be compared with the discussion about the β_i terms in section 4.3.2 on page 124. By up-scaling the integrators unit gain frequency or the feedback β-factors towards the output, the zeros formed by the short circuiting β-paths cancel the influence of all but the last integrator on the limit cycle oscillation. Since the contribution of the last integrator is a 90° phase shift, its GBW can be placed as high as possible, improving the linearity. This effect is shown in figure 4.36. By a simple alteration of the integrators unit gains, the third order harmonic distortion improves with almost 10 dB, leaving the maximum below the zeroth order SOPA limit.

3 In a third region, the distortion decreases again. This is not due to an increase in linearity, but to a loss in forward gain. Above the unit gain frequencies of the integrators, the closed loop gain decreases. This can be predicted from equation (4.113). As a consequence, the error signal B also decreases. The third order distortion, which is proportional with the third power of B, thus decreases faster than the output voltage decay. Hence the HD3-number improves. The big drawback of using a SOPA in this region

is the extra signal attenuation in the high frequent regions. For some extent this can be compensated in the digital domain, by the channel equalisation filters that are inherent to an xDSL-system. But this is a system wide design decision, and will not be further elaborated in this work.

3.4.2 Signal Bandwidth

Like for the zeroth order case (see section 4.2.3.3), the signal bandwidth will be determined by the additional effect of two degradation mechanisms : the sideband modulation bands and the time invariance consideration.

It has been shown that the Bessel series approximation of the modulation bands that has been performed in section 4.2.3.3, becomes more accurate for more $\Delta\Sigma$−like configurations [Roza, 1997]. The sideband modulation terms will thus be given by equation (4.92). It has already been shown that for coupled higher order SOPAs, the common mode oscillation is the dominant oscillatory mode. So all odd harmonic modulation bands will be cancelled. For most practical cases this means that the signal bandwidth is limited by the $2 \times (\omega_{LC} - \omega_{sig})$ frequency peak. In this, ω_{LC} denotes the limit cycle frequency and ω_{sig} the maximum signal frequency.

Concerning the second bandwidth limitation, being the time invariance criterion of the describing function technique, it was generally noted by equation (4.75) that this degradation effect can be regarded as an extra distortion term added at the output of the comparator. In order to have the time invariance term to have an equal importance as the distortion term due to inherent non-linearity, equation (4.77) is still the basic bandwidth limitation. However, the distortion introduced by the time invariance criterion is also shaped by the integrators in the forward path. Compared with a zeroth order SOPA however, the bandwidth for which a certain distortion level can be guaranteed can be higher for a higher order SOPA, dependent in which zone of the HD3-curve of figure 4.35 the $\omega_{LC}/3$ bandwidth is positioned.

This effect is illustrated in figure 4.37. These results were obtained by numerical simulations of two SOPA amplifiers for different input signal frequencies. The filter cut-off frequencies are positioned in such a way that the limit cycle frequency is the same for both amplifiers. This to obtain a fair comparison. The integrator GBWs were positioned at 1 MHz 4 MHz 16 MHz. Both SOPAs were coupled with a coupling factor α of 0.05, in order to solely measure the time invariance induced degradation and not the modulation sidebands. The signal amplitude was set to 0.1 V.

As can be clearly observed, the higher order SOPA virtually has a higher bandwidth as the zeroth order amplifier, even when the maximum of the HD3 curve is higher than the normal HD3 number of a zeroth order amplifier. The use of a coupled higher order SOPA amplifier relaxes thus the bandwidth to the second order modulation term.

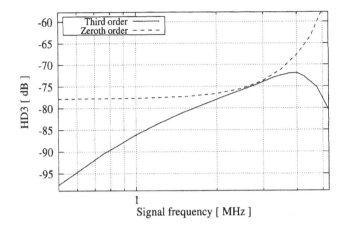

Figure 4.37: Comparison of the third order distortion at the systems bandwidth for a zeroth and a third order SOPA

4.   Final Remarks and Conclusions

In this chapter a reference Self Oscillating Power Amplifier has been introduced. Generally speaking a SOPA is defined as the coupling of two basic SOPA building blocks by a line transformer / line impedance combination. Each SOPA building block is a feedback loop consisting of zero or more integrators followed by a comparator. The loop is closed using a loop filter towards the input. Additional loops are constructed by β_i amplification blocks towards the inputs of every integrator. A SOPA amplifier is called an nth order SOPA with n the number of integrators in the loop.

Since a SOPA is a hard non-linear circuit, numerical simulations are very time-consuming and linear control theory cannot be applied to gain insight in the systems performance. In this chapter a complete analysis of the SOPA system has been presented, based on the theory of the describing function analysis. From this analysis the following conclusions can be drawn :

- An autonomous zeroth order SOPA system will oscillate with a fixed frequency if the order of the loop filter is three or higher. For a higher order SOPA, β_i local feedback factors are used to stabilise this oscillation. Ideally this limit cycle oscillation and frequency is determined by the loop filters cut-off frequency and order.

- In a coupled SOPA system it has been shown, that for all realistic line conditions and thus coupling factors between both oscillators, oscillator pulling towards synchronisation always occurs. In this way the limit cycle oscillation becomes common mode with respect to the primary coil of the line transformer. This oscillation is thus completely cancelled towards the line. From a system performance point-of-view, this effect will increase the power amplifiers efficiency and reduces the out-of-band filtering requirements.

- The limit cycle oscillation will act as a natural dither in the SOPA system. This effect is shown by calculating the distortion of a forced oscillation. By the linearising effect of the limit cycle, a very linear modulation can be obtained without high OSRs. This is due to the fact that unlike in synchronous switching systems, the input can be represented by a continuous time, discrete amplitude signal, shifting the amplitude accuracy issues to time domain accuracy.

- Another side-effect of the natural dither is the inherent adaptivity of the system towards circuit non-idealities. This was shown by calculating the example of a comparator with finite gain.

- It is shown that a SOPA system is able to meet the linearity specs for ADSL as long as the noise level of the circuits can be kept low enough. This noise limit can be relaxed by either going to technologies (or circuit techniques) that are able to handle high supply voltages or to design higher order SOPAs to lower the in-band distortion. By the introduction of integrators in the forward path a noise-shaping technique is introduced in the system.

- A thorough bandwidth analysis has been performed and two major contributors were defined : the effect of the time-variance properties of the driven system and the modulation sidebands around the harmonics of the limit cycle frequency. It was shown that all sidebands around the odd harmonics of the limit cycle frequency cancel out and that the limit is given by the time-invariance consideration, which is a value of one third of the limit cycle frequency. This limit can be elevated by going to higher order SOPA systems.

- By the calculation of the sideband powers, the completed model explains every peak in the spectrum of the SOPA's output.

- All calculated models have been compared with numerical simulations. It is shown that the models correspond with the simulated behaviour. The error margins are that small that the presented analysis can be used to study and design the complete amplifier on the system level. In the next chapter the reduction of design time using the presented model will be further discussed.

- Although the presented methodology is focused on the design of a SOPA amplifier as defined in the first section of this chapter, the analysis systematics can be applied to all analogue non-linear systems as long as the filter criterion holds in the loop transfer function.

Chapter 5

HIGH-LEVEL DESIGN PLAN SYNTHESIS AND CAD-TECHNIQUES FOR NON-LINEAR SOPA DESIGN

IN the previous chapter, a complete model for the SOPA amplifier has been derived. This model not only provides insight in the design of a SOPA, but also a very fast and accurate tool to explore the design space from the systems point-of-view. In this chapter the obtained models will be used to construct a design plan for a SOPA amplifier starting from the specifications towards the circuit characteristics. The design plan in that way is the bridge between the high level modelling of previous chapter and the actual circuit design of next chapter. To accomplish this linking function, the knowledge obtained by the modelling effort is combined with basic circuit knowledge without going into too much detail.

This design methodology, however, cannot be followed without a set of design tools that support a systematic design. The general specifications for these CAD-tools will be discussed. A set of tools that has been written for this specific task will be presented. The main goal of these pieces of software is to provide the designer with a consistent and accurate way to design a SOPA system with a minimum of design iterations.

The chapter is organised as follows : firstly the design limitations are derived from the technology limitations. For this the scaling laws of mainstream CMOS technologies are converted into an estimation of the upper limit of the specifications a SOPA can reach. Next the basic properties of a line transformer are explained. This consideration with the choice of process technology provide the basic feasibility study of the SOPA as a power amplifier for a certain application. The rest of the design plan synthesis section will cover every step in more details. A power estimator to determine the maximal power efficiency closes this section.

The other half of this chapter is dedicated to the composition of a set of CAD-tools. The emphasis is put on the requirements of a tool set, not on

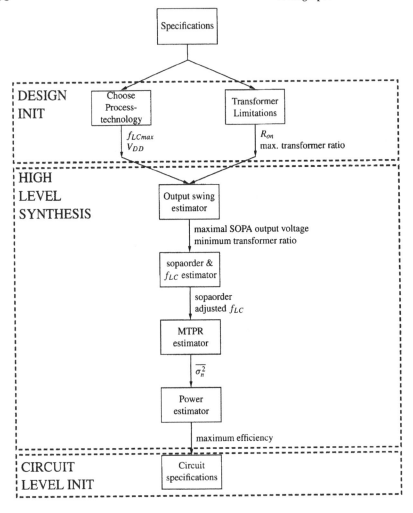

Figure 5.1: Schematic overview of the design plan

the actual implementation. However, code snippets are extensively used to illustrate the most important coding ideas. The tool set is divided into four parts : the behavioural modelling, the numerical simulation, the model library and the measurement interface. Each part will be elaborated in separate sections, with an emphasis on the mathematical properties and their influence on the design flow.

1. Design plan synthesis

In figure 5.1 a schematic overview of the complete high-level design plan is given, without iteration paths. Although the design plan is devised in order to eliminate these iteration steps as much as possible, they cannot be avoided completely. So from every point in the schematic of figure 5.1, there should be a path back up to every preceding block.

In the construction of this design plan, three distinct zones are marked:

The Design Initialisation is defined as the set of design decisions that are very important for the conception of the SOPA amplifier but are not a real part of the system. The most important examples of such decisions are the process technology choice and the transformer design/choice. Design decisions taken in this part of the design will transform the imposed specifications into more implementation oriented specifications.

The High Level Synthesis layer deals with the mathematical models obtained from the previous chapter. In this step the complete system level architecture is calculated without the knowledge of the actual implementation. The models, however, utilise the process parameters and have estimators to calculate a first order implementation feasibility. Since these calculations are based on the behavioural models from the previous chapter, the system exploration is a fast step. Due to the inherent adaptivity, see section 4.2.5, an architecture that still has some margin to fulfils the requested specifications should be implementable.

The Circuit Level Initialisation step is the layer where the specifications for different building blocks are derived. As inputs this step takes the complete system level architecture, combined with the knowledge of the basic limitations of possible building block implementations. Since the setup of a behavioural SOPA model that incorporates most of the higher order non-idealities of real circuit design, this step mostly depends on numerical simulations.

Note that this design plan is overall a top-down methodology, but by incorporating the feedback from the circuit level initialisation layer, a bottom-up implementation path is available. When the a priori knowledge of possible implementations is not available, it is good design practice to incorporate some safety margins in the following calculations. These margins can be set narrower when the implementation feedback path raises the design knowledge level. In what follows, every block of figure 5.1 is further elaborated.

1.1 The Design Initialisation Layer

1.1.1 Limitations imposed by the processing technology

Eventually the amplifier needs to be realised in a certain technology. The parameters of this technology will impose the basic limits to the design of the SOPA. Since a complete design a the borders of the state-of-the-art is very

Table 5.1: CMOS scaling laws following the SIA Roadmap

Parameter	Relation	Full Scaling	General Scaling	Fixed-voltage Scaling
W, L, t_{ox}		$1/S$	$1/S$	$1/S$
V_{DD}, V_T		$1/S$	$1/U$	1
N_{SUB}	V/W_{depl}^2	S	S^2/U	S^2
C_{ox}	$1/t_{ox}$	S	S	S
C_{gate}	$C_{ox}WL$	$1/S$	$1/S$	$1/S$
β	$\mu C_{ox}W/L$	S	S	S
f_T	g_m/C_{ox}	S^2	S^2	S^2

tedious, it is of an utmost importance that the technological limits can be determined as early as possible in the design process, preferably before actual circuit design has started. If the designer has access to various technologies, the right technology choice from the start, will speed up the complete design. Of course, the technology parameters will influence every component in the system, so a strict determination of the possibilities of a certain technology is not possible. The considerations presented here, have to be regarded as rules-of-thumb that give strong indications for the feasibility of the requested specifications. In this discussion, we will limit ourselves to CMOS technologies. The CMOS scaling laws were given in table 5.1 [Iwai, 1999, Rabaey, 2002]. Three different scaling laws are considered : the fixed-voltage scaling which was the dominant scaling law for gate lengths longer than 0.7 μm. Going to smaller dimensions, the supply voltage needed to scale with the gate thickness, to avoid oxide breakthrough. For submicron technologies full scaling needs to be applied. The middle column gives a model that can be used in both regions by setting the voltage scaling factor U. To calculate the applicability of a technology to construct a SOPA, the assumption is made that the process will limit the maximum switching frequency. The abstraction of noise and distortion considerations as presented in figure 4.20 is made since these specifications can be elevated by going to higher order SOPAs.

Since the SOPA basically is a power amplifier, the most important technology specification is the maximal applicable supply voltage[1]. This supply voltage will set the necessary transformer ratio, since the output levels are related to the output power specification and the line impedance. In this way the size of the output driving transistors is limited by the requested output power. If the

---

[1]In this work, techniques that increase the maximum supply voltage for a low-voltage technology like drain-source engineering [Annema et al., , Sowlati and Leenaerts, 2002] will not be considered.

base inverter x times x^2 times x^n times

Figure 5.2: Schematic representation of a tapered buffer

maximum output swing of the SOPA is assumed to be one fifth of the supply-voltage, the necessary output resistance R_{on} for a necessary output power P_{out} and an output efficiency of at least 90% should be :

$$R_{on} = \frac{0.1V_{DD}^2}{25P_{out}} \sim 1/U^2 \qquad (5.1)$$

From (5.1), it can be derived that the on-resistance specifications for the output stage become more severe with transistor scaling. Since the on resistance is given by :

$$R_{on} = \left(K_P \frac{W}{L} \left(V_{GS} - V_T - \frac{V_{DS}}{2} \right) \right)^{-1} \sim \frac{U}{S^2 W} \qquad (5.2)$$

the width of the output transistor W, will thus scale with U^3/S^2. This means that the total transistor area scales with U^3/S^3. This output transistor is driven by a tapered buffer as shown in figure 5.2. The ratio between the input capacitance of the standard inverter and the input capacitance of the output driver y can be be calculated as :

$$y = \frac{W}{W_{min}} = \frac{25L_{min}P_{out}}{0.1W_{min}V_{DD}^2 K_P(V_{DD} - V_T - 0.05V_{DD})} \sim \frac{U^3}{S^3} \qquad (5.3)$$

Using this relationship the minimal time delay of the tapered buffer can be calculated [Rabaey, 1996]:

$$t_{dmin} = (e \ln(y)) t_{d0} \qquad (5.4)$$

with t_{d0} denoting the delay of a basic inverter. An approximation for this delay is given by :

$$t_{d0} = \frac{24L_{min}^2 C_{ox}}{K_P V_{DD}} \sim \frac{U}{S^2} \qquad (5.5)$$

For the limit cycle to be independent of the delay of this digital buffer, a rule-of-thumb is that this delay should be lower than 10% of the limit cycle period. The highest reachable limit cycle frequency is thus given by :

$$f_{LCmax} = \frac{0.1}{t_{dmin}} \sim \frac{S^2}{U(C^t + 3\ln(U/S))} \qquad (5.6)$$

For a .35 μm CMOS technology, this limit is calculated to be 37 MHz. Since the bandwidth of the amplifier is maximum one third of the limit cycles frequency (see section 4.2.3.3), the maximum attainable bandwidth in a .35 μm CMOS technology will thus be around 12 MHz which is sufficient for VDSL downstream signals. From (5.6) one could derive that when it is possible to use the fixed voltage scaling laws, the bandwidth in first order goes up with the square of the scaling factor. If the specifications are more stringent and one needs to use technologies below .5 μm, so with voltage scaling, the bandwidth scales in first order proportional with the scaling factor. However other effects like transformer ratio and noise limitations will degrade this scaling very rapidly. Moreover, going to deep sub-micron technologies will risk a rapid degradation of the efficiency due to the increased leakage current. With other words : scaling down below the limit where scaling laws are dominated by full scaling rather than fixed-voltage scaling will not raise the SOPAs specifications. Solutions to overcome this limitations are e.g. :

■ most novel sub-micron technologies provide 5 V output transistor to be compatible with older digital building blocks. These transistor will not be as fast as their regular counterparts in those technology. From the design of a SOPA point-of-view, these transistors extend the fixed voltage scaling region.

■ the use of more specialised technologies like Bipolar assisted CMOS (BiC-MOS) or Diffusion Metal-Oxide-Semiconductor (DMOS) transistors [Zojer et al., 1997]

■ the use of techniques that enable higher voltages in low-voltage technologies, like drain-source engineering

■ in this analysis some specifications were put in as fixed values, like a 90% output efficiency. Toying with these specification can enable a SOPA design if the required specifications come close to the presented technological limits.

1.1.2   Transformer Limitations

From figure 4.20, one could falsely conclude that in theory every possible distortion level can be reached, the output noise-level could be set low enough. However the construction of a high-bandwidth transformer with a high transformer ratio could be quite cumbersome. This subsection will reflect on the limitations imposed by the line transformer, without going into too much details, since the design of a line transformer [Dixon, 1999] for xDSL is beyond the scope of this work.

The transformer ratio at the high bandwidth of an xDSL system is limited by the required output efficiency. Since the output impedance R_{line} is transformed into R_{line}/n^2 for a transformer with a transformer ratio n, the parasitic series resistance of the primary winding will become dominant.

$$\epsilon = \frac{R_{line}}{R_{line} + n^2 R_p} \tag{5.7}$$

The parasitic series resistance will always be high for a high bandwidth line transformer compared with regular transformer design, since :

- The skin effect will force the current to only flow through the perimeter of the wire. The penetration depth (D_{PEN}) is inverse proportional with the square root of the frequency :

$$D_{PEN} = \sqrt{\frac{\rho}{\pi \mu_0 \mu_r f}} \tag{5.8}$$

  This will heavily increase the parasitic resistance of one winding. The perimeter could be solved by putting more windings in parallel. However the magnetic interaction between parallel windings in different layers drives the current out of the tangent planes, increasing the parasitic resistance on its turn.

- The number of parallel windings, however, is limited by the core losses at high frequencies. To meet the bandwidth specifications, a ferrite NiZn-core will be the most popular to use. High frequency cores have a higher resistivity to minimise eddy current loss. The number of windings will then be limited by the saturation flux density.

- Another limit on the number of turns is the inter-winding capacitance and parasitic spread inductance. These parasitic elements grow proportional with the square of the number of turns and limit the transformers bandwidth, since they introduce a second order low-pass filtering at the input.

- The size of the core should be kept small to lower the impact of the core losses. The following rule-of-thumb indicates an estimate for the area-product, which is the product of the area of the window times the cross section of the magnetic core and thus a good measure for the volume of the transformer, of a line transformer:

$$AP \quad = \quad A_W A_E = \left(\frac{P_O}{K \Delta B \ f_T}\right)^{\frac{4}{3}} \ cm^4 \qquad (5.9)$$

with :

A_W the window area

A_E the magnetic cross section

P_O output power

ΔB flux density swing

f_T transformer frequency

$K = .017$ for full bridge converters

Note that the transformer frequency is not necessarily the same as the maximum input frequency. For instance, for push-pull output stages the transformer frequency is half the input frequency. The maximum number of turns is thus also limited by mechanical limitations.

■ After selecting the core material and shape, the number of turns can be calculated by using Faraday's Law at the primary of the transformer, since this site will contain the smallest number of turns.

$$N_P = \frac{V_{in}}{f_T \Delta B A_E} \qquad (5.10)$$

To minimise the overall losses, the number of turns should be of integral value. This will also limit the transformer ratio.

All practical considerations taken, a maximum transformer ratio of 7 to 8 is possible. Higher values should not be considered.

1.2   High Level Synthesis Layer

1.2.1   Output Swing Estimation

The maximum output swing can be estimated as the specified rms output voltage divided by the transformer ratio and multiplied by the CF of the input signal. The minimal transformer ratio can be calculated by turning this reasoning the other way around :

$$N_{min} \simeq \frac{Vout_{RMS} \ CF}{V_{DD}} \qquad (5.11)$$

A first feasibility check is to check this minimal ratio with the obtained maximal ratio. If the obtained transformer ratio is not feasible another process technology that can cope with higher supply voltages need to be chosen or the requested specification should be discussed.

The interval between the minimal and maximal transformer ratio can be used as a degree of freedom to reach the distortion specifications of the amplifier. To calculate the minimal third order distortion, the graph of figure 4.18 is constructed using the distortion analysis of section 4.2.3.2. Using the calculated minimal and maximal output swing, the linearity limits can be determined. The results of this analysis are taken towards the next design step.

1.2.2 SOPA order estimation

If the obtained linearity interval for the zeroth order does not fulfil the specifications, a shift towards a higher order SOPA has to be made. An important distinction can be made between systems with flat linearity specifications and frequency plans with lower bit-loading at higher frequencies like the VDSL CO deployment scenario mask [Wang, 2001] :

Flat distortion specifications in the whole bandwidth, need to
be met by suppressing the distortion of a zeroth order SOPA ($HD3_0$) with extra integrators in the forward path as illustrated in figure 4.35. By properly choosing the feedback factors (β), the distortion attenuation can start at approximately one fourth of the limit cycle frequency. In this way the requested distortion level connects the limit cycle frequency with the order m of the SOPA amplifier :

$$HD3_{spec} - HD3_0 = \frac{m \times 20 \text{ dB} f_{LC}}{4 \, BW} \qquad (5.12)$$

The above relation is limited by the maximal limit cycle frequency f_{LCmax} from the chosen process technology. So the minimum SOPA's order can be calculated. Note that m can only take integer values. So a set of solutions is obtained from which the lowest orders are preferred to limit the necessary device area. However a lower mean switching frequency allows higher efficiencies as long as the added power consumption of an additional integrator is lower than the decrease in dynamic power consumption of the output drivers.

For a shaped distortion spec, a lower power consumption can be obtained if the loop filter follows the shape of the power density mask. The goal is to create an extra gain in the loop filter so that the limit cycle frequency and the number of integrators in the forward path is minimal.

From this analysis a limited set of possible SOPA architectures follows. The architectures are characterised by the SOPA order, the limit cycle frequency, the unit gain frequency of each integrator and the according β_i factors.

1.2.3   MTPR estimation

For every architecture the MTPR can be easily calculated using (4.67), so a graph like the one of figure 4.20 is obtained. From this graph the maximally allowed noise density can be derived. This noise level will determine the power consumption in the first integrator, since this will dominate the overall noise limits. The noise generated by the other integrators will be shaped by the preceeding integrators in the loop. If the designer has some knowledge of obtainable noise levels, a selected set of architectures can be omitted in this step.

1.2.4   Power estimation

To estimate the total power consumption of the SOPA, the following contributors should be considered :

The dissipation in the output transistors should be the dominant contributor. In the determination of the process technology, it was set to 90% of the output power. This value is also taken in this step to determine the output resistance and the input capacitance of the output transistors.

The power dissipated in the output drivers can be estimated as

$$P_{driver} = 2\,V_{DD}^2\,f_{LC}C_{gate}(x + x^2 + x^3 + \cdots + x^n) \qquad (5.13)$$

$$= 2\,V_{DD}^2\,f_{LC}C_{gate}x\frac{x^n - 1}{x - 1} \qquad (5.14)$$

$$\simeq 2\,V_{DD}^2\,f_{LC}C_{gate}\frac{y}{e - 1} \qquad (5.15)$$

y is the ratio of the output drivers input capacitance and the gate capacitance C_{gate} of a minimal base inverter.

The power consumption of the integrators can be estimated from the required noise levels. If a fraction of the total noise budget is reserved for the integrators iNF_{int}, the integrated noise over the bandwidth BW can be expressed as :

$$\frac{kT}{C_{int}} = BW\,\overline{\sigma_n^2}\,NF_{int} \qquad (5.16)$$

Since the unit gain frequency of the first integrator is around the systems bandwidth, the following relation holds for a $g_m - C$ implementation:

$$GBW_{int} = \frac{g_m}{2\pi C_{int}} = \frac{2I_{DS}}{2\pi C_{int}(V_{GS} - V_T)} \simeq BW \qquad (5.17)$$

The bias current of the complete integrator can be approximated as four[2] times the necessary current to obtain the wanted transconductance. The power consumption can thus be estimated as :

$$P_{int} = 4V_{DD}I_{DS} \tag{5.18}$$

$$= 4V_{DD}(V_{GS} - V_T)C_{int}BW \tag{5.19}$$

$$= \frac{4V_{DD}(V_{GS} - V_T)kT}{\overline{\sigma_n^2}\,\mathrm{NF}_{int}} \tag{5.20}$$

The power consumption of the integrators is thus independent of the bandwidth of the system, but dictated by the noise budget of the system. The complete power consumption can be in first order estimated by multiplying the integrators power consumption with the number of integrators. This is a safe over-estimation since it is possible to scale the integrators towards the comparator.

The power consumption in the feedback filter can be calculated in the same way. If a portion NF_{RC} of the noise budget is spendable in the loop filter, then the following relation holds for a RC - loop filter:

$$\overline{\sigma_n^2}\,\mathrm{NF}_{RC} = 4kT\,(nR) \tag{5.21}$$

The power consumption of this loop filter is calculated by considering the voltage drop of the limit cycle oscillation over the resistors of the nth order loop filter with cut-off frequency f_C :

$$P_{RC} = \frac{V_{DROP}^2}{R} \tag{5.22}$$

$$= \frac{V_{DD}^2}{R}\frac{\left(f_{LC}^2 + f_C^2\right)^n + f_C^{2n} - 2f_C^n\sqrt{\left(f_C^2 + f_{LC}^2\right)^n}}{\left(f_C^2 + f_{LC}^2\right)^n} \tag{5.23}$$

$$= \frac{4V_{DD}^2kTn}{\overline{\sigma_n^2}\,\mathrm{NF}_{RC}}\frac{\left(f_{LC}^2 + f_C^2\right)^n + f_C^{2n} - 2f_C^n\sqrt{\left(f_C^2 + f_{LC}^2\right)^n}}{\left(f_C^2 + f_{LC}^2\right)^n} \tag{5.24}$$

With f_{LC} given by (4.19) respectively for a zeroth order SOPA (4.118) for a higher order SOPA, it can be proven that the power consumption in the loop filter decreases with increasing filter order.

Adding these power contributions together will give an estimate for the achievable efficiency of the line driver. So the architecture from the restricted set out of the preceding steps with the highest efficiency can be withheld.

---

[2]Since the integrators have four inputs, if a fully differential implementation is chosen

1.3   Circuit Level Initialisation

By finishing the previous steps, the architecture of the SOPA-under-design is fixed. The circuit level initialisation step can be regarded as a preparation step towards the complete analogue implementation. Some more parameters of the different building blocks are derived in this step. A short but incomplete list of possible parameters are :

- The values of the β_i factors and the integrators unit gain frequencies follow straightforward from the architectures specifications. The minimal capacitance to implement this building block follows from the noise consideration of (5.16).

- The finite gain of the integrator has been integrated in the models of (4.4). The minimal gain can be easily obtained from numerical simulations.

- The distortion of the first integrator will be completely visible at the output. Design for low distortion will govern the design of the first stage. Design limits for the other integrators can be easily derived from additional numerical simulations.

- The output transistors are governed by the total power consumption, the transformer ratio and the supply voltage (5.1). From the technology parameters, the total input capacitance can be calculated.

- The digital buffer to drive this capacitance is designed for minimal delay [Rabaey, 1996].

- The design of the loop filter is also governed by noise considerations (5.21). Together with the values of the corner frequencies, which can be easily derived from the limit cycle frequency, the complete design of the loop filter is determined.

- The finite gain of the comparator has been modelled in section 4.2.5. These models set limits on this circuit specification.

- In the same way comparator hysteresis can be modelled [Gelb and Vander Velde, 1968] and appropriate limits can be set.

2.   CAD-tools to support the design methodology

2.1   Requirements

As been explained in section 4.1.3, numerical simulations, even of the simplest mathematical model is very time consuming due to the frequency content of the modulation. To optimise the design speed, a limited set of design tools is necessary on top of the 'classical' circuit simulators. In this work, an example of an implementation of a set of tools is sketched. The basic requirements for the design of this bundle of tools was :

- The tool set should be easily extensible to new architectures without rewriting too much code. In other words, it should provide a limited prototype-platform.

- The set should cover the complete design plan, from the theoretical models of chapter 4 to the circuit level implementation within the same framework.

- It should be possible to regard the design on different abstraction levels with the same parameter set. This means the construction of interfaces between the different abstraction levels.

- It should be possible to speed up simulations of fixed architectures, i.e. architectures that have passed the prototype level.

- It should be fairly easy to introduce non-idealities at different stages of the design in order to check their influence on the complete system behaviour without too many lengthy simulations.

A true CAD-tool should also consider the user interface, but this lies beyond the scope of this text.

2.2 The Octave framework

2.2.1 Framework Overview

Figure 5.3 shows a schematic overview of the implemented design framework. In the centre the octave mathematical computing language [Eaton, 2002] has been chosen as the heart of the framework. Octave is a MATLAB-like computing engine. The main advantages of the octave language are:

- It's compatibility with MATLAB opens a wide variety of numerical helper functions.

- Octave is open-source software. Next to the fact that one saves the license cost of the software, the use of open-source software for numerical computations adds the following benefits :

  - The numerical engines are peer-reviewed, mostly Fortran, implementations taken from netlib (http://www.netlib.org) who are written by top mathematicians. By studying the implementations, the designer of the numerical simulators can learn the difficulties about the implementation of numerical algorithms and possible solutions.

  - Octave comes with a C++ library which can be used to implement heavily used functions in C, C++ or Fortran. These functions can be loaded

Numerical Engines

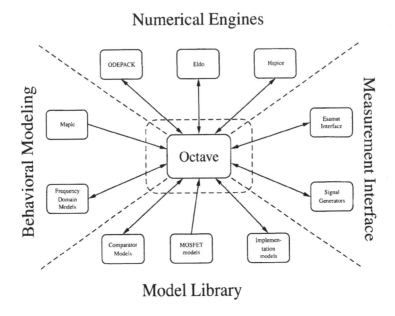

Figure 5.3: Schematic overview of the design framework

dynamically in the octave environment. Due to the open-source nature, many examples of dynamically loadable functions can be examined and adapted.

- Octave has built-in functions for visualisation, and system interface like Unix-pipes which saves design time.

- Octave has a built-in control theory toolbox, which enables the construction of a block diagram via a textual input language. A textual input has the advantage that it provides a more flexible way to automate the creation of different architectures than schematic entry tools like e.g. simulink.

Around this octave engine several applications are connected by custom interfaces. These are presented by the arrows in figure 5.3, in which the two-sided arrows represent building blocks for which a two-way communication has been developed. The applications are grouped in four separate clusters:

The behavioural modelling cluster combines the describing function models derived in chapter 4 with the symbolic calculus engine of the MAPLE program. The goal of this cluster is to create behavioural models for a wide variety of architectures. The outcome can be in the form of closed expressions or octave-functions which can be easily evaluated in the frequency domain.

The numerical engines consist mainly of the ODEPACK differential equation solver and the industrial circuit simulators ELDO and HPSICE. The ODEPACK-simulator has been extensively used in the previous chapter and has been described in section 4.1.3. The interfaces with both circuit simulators will be described more extensively in the following of this chapter. There is a possibility to spread the necessary processing power over a number of computers. This can speed up calculations for heavy numerical calculations. The results are time domain signals which need to be post-processed by octave.

The model library contains the different models who have been derived in the behavioural modelling cluster and the numerical models for the different architectures. An important part of the library is a set of parameterisable implementation architectures for different building blocks and of course the models for the used process technology.

An interface with the measurement setup has also been provided. This consists of a signal generating part, which can also be used in the numerical simulators and a part which is able to visualise and post-process measurement data. The integration in the complete design system enables a direct control of the measurements with the results form the modelling effort.

2.2.2   Parameter structures

In order to be able to step flexibly between the different abstraction levels, the concept of parameter structures has been introduced. A parameter structure is an abstraction level. It is constructed as an object that contains all necessary parameters to describe the SOPA architecture and a set of functions to translate this parameter set into the necessary instructions for the different abstraction layers. The parameter structures can on its turn contain references to other parameter structures.

For example the SOPA parameter set of a higher order SOPA will contain the different integrator unit gain factors. Since the Maple model approximates the architecture so that all integrators have the same integrator frequency, the model only takes the first parameter. The octave implementation of the frequency domain models can handle different integrator GBWs, so the complete set is used. In the SOPA parameter set there can be a link towards an integrator parameter set that is coupled with a certain implementation, providing functions that in first order translate the values of the SOPA parameter set into appropriate g_m and C values and so on towards W and L values of the transistor implementation.

2.3   Behavioural Modelling

2.3.1   Describing function analysis using Maple

For simple architectures like e.g. a zeroth order SOPA with equal corner frequencies, the use of a full symbolic analysis has the benefit that it produces closed expressions from which more information can be derived. In order to generate human-readable expressions, several mathematical considerations need to be kept in mind :

- The *arctan* - command implements a mathematical inverse tangent, which defines output values between $-\pi/2$ and $\pi/2$. As a consequence, a phase shift over $\pi/2$ causes a wrap around in the mathematical model and can thus lead to incorrect simplifications or
  tan(arctan(x))-constructs that cannot be simplified automatically in the Maple-scripts.

- To calculate only the real root of a cubic equation, like in the distortion model, Vièta's substitution is used.

- The *series*-command needs to be heavily used to create linear approximations of the obtained expressions.

- Substitutions need to be performed as late as possible, since a simplification step of a large expression slows down the computations.

The maple program provides a Fortran and C interface, so that results can be introduced in the framework. Also a file interface, using numerical evaluations in maple can be used.

2.3.2   Numerical implementation of the describing function model

In order to accommodate the problems with the problems of the symbolic analysis using maple, the algorithms have been implemented in the octave numerical computing language.

In code snippet 5.1 the code for the behavioural calculation of the limit cycle frequency is shown. The limit cycle frequency is using the function *fsolve*, which is based on the MINPACK [Powel, 1970] subroutine *hybrd*. As a starting value the limit cycle frequency for a first order SOPA is taken. As it has been shown, the goal of the β_i feedback stages is to introduce zeros into the loops transfer function. So the first order approximation is an excellent initial value for the nonlinear equation solver. An accurate value for the limit cycle frequency and amplitude is obtained without the inverse tangent limitations. The outcome does not depend on numerical simulations, so the evaluation is very fast. Listing 5.1 shows the octave implementation of a parameter structure. The global variable *sopaparam* contains all data to describe the SOPA architecture.

Listing 5.1. Behavioural calculations of the limit cycle frequency

```
function  out=g(omega)
        global  sopaparam;
        tfint=0;
        for  teller=1:sopaparam.sopaorder,
                tfint=(tfint+sopaparam.beta( ...
                teller))
                        *sopaparam.fint/(I*omega) ...
                        ;
        endfor
        out=sopaparam.filterorder*atan(omega/ ...
        sopaparam.ffil)
                -pi-angle(tfint);
endfunction

function  [varargout]=limitcycle_calcul(varargin)
        global  sopaparam;

...Parameter sweep code snipped

        startvalue = sopaparam.ffil*tan(pi/2/ ...
        sopaparam.filterorder);
        [hulp,INFO,MSG]=fsolve("g",startvalue);

...Post-processing and error control code snipped
endfunction
```

In the same way, the distortion analysis of figure 4.34 has been implemented in code snippet 5.2. The structure of the code follows the structure of the algorithm as closely as possible to enable qualitative evaluations of the model. The solution of the cubic equation left aside, the code does not contain iteration. Notice also the use of the parameter structure *signalparam*. It is referred by the signal-field in the sopaparam structure. Also the limit cycle amplitude and oscillation is stored in the *sopaparam* - structure, since it characterises the SOPA from a behavioural point-of-view. These values can be used in a numerical simulation to determine a priori, the simulators step -and stop times.

Listing 5.2.   Distortion calculation using the behavioural model

```
function out=distogencal(B)
        global sopaparam;
        signalparam=sopaparam.signal;
        # limit cycle values
... Code to check of recalculation if limit-cycle is necessary, is snipped
        omega=sopaparam.omega;
        A=sopaparam.A1;
        # forward (G) and feedback (L) filter ...
            generation
... L and G generation code snipped
        # cubic equation to calculate error- ...
            signal B
        NB1=sopaparam.VDD/pi/A;
        NB3=sopaparam.VDD/pi/A^3/8;
        b3=3/4*NB3*G*L;
        b2=0;
        b1=NB1*G*L+1;
        b0=signalparam.amplitude*G;
        out=b3*B^3+b2*B^2+b1*B+b0;
endfunction

function [varargout]=evaldisto()
        global sopaparam;
... L and G generation, and describing function definition code snipped
        [distogencalval ,INFO,MSG]=fsolve(" ...
            distogencal",signalparam.amplitude/2);
        vr_val(20*log10(abs(distogencalval) ...
            ^3*1/4*NB3/(1+NB1*L*G)));
endfunction
```

2.3.3   Graphical frequency domain analysis

The octave control systems toolbox provides functions to define control systems in a block diagram kind of way. System building blocks can be defined by their state-space (ss2sys), pole-zero (*zp2sys*) or transfer function (*tf2sys*) representation. Various commands are also provided to connect this building blocks into a complete system. Listing 5.3 shows an example of the construction of a SOPA-model in the frequency domain.

Listing 5.3. Behavioural frequency domain model using the octave's control systems toolbox

```
function sopa_model=laplace_model()
    global sopaparam;

    integrator=ss2sys([0],
                      [2*pi*sopaparam.fint -2* ...
                       pi*sopaparam.fint],
                      [1],[0 0]);
    filter=ss2sys([-2*pi*sopaparam.ffil],[2* ...
        pi*sopaparam.ffil],
                      [1],[0]);
    H = filter;
    for i=1:(sopaparam.filterorder-1),
        H= sysmult(H,filter);
    end
    sopaparam.H = H;

...Code to define the integrators + local feedback snipped

    sopa_model=sysmult(H,G);
endfunction
```

Using these models, most graphical methods from control theory can be performed like Bode, Nyquist, or Modified Nyquist plots like the ones shown in figure 4.6. As in the behavioural modelling simulator of previous subsection, the graphical methods use the *sopaparam* parameter structure. Since the frequency domain analysis use the same input parameters, but complete different algorithms, they can be used to control the results of the behavioural modelling and vice-versa.

The graphical methods provide a direct insight in the behaviour of the SOPA amplifier. They provide a direct interpretation for the design of a loop filter for a system with a non-flat power spectrum density mask.

2.4 Numerical simulations

2.4.1 ODEPACK implementation

The state-space equations for a SOPA amplifier have been derived in section 4.1. Since the state-space equation show a remarkable symmetry (4.4), the system equations can be very flexibly created by the addition of diagonal matrices. Listing 5.4 shows as an example the creation of the state-space *A*

Listing 5.4.   Creation of the system equations for the ODEPACK simulator

```
A1=diag(2*pi*sopaparam.ffil*ones(1, ...
    sopaparam.filterorder −1),−1)+diag(−2* ...
    pi*sopaparam.ffil*ones(1,sopaparam. ...
    filterorder));
A2=zeros(sopaparam.filterorder,sopaparam. ...
    sopaorder);
A3=A2';
A4=diag(2*pi*sopaparam.fint*ones(1, ...
    sopaparam.sopaorder −1),−1)+(sopaparam. ...
    finite_gain)*diag(−2*pi*sopaparam.fint ...
    *ones(1,sopaparam.sopaorder)/sopaparam ...
    .Aint);
for i=1:sopaparam.sopaorder,
        A3(i,sopaparam.filterorder)=−2*pi ...
            *sopaparam.beta(i)*sopaparam. ...
            fint;
endfor
A=[A1 A2 ; A3 A4];
```

matrix. The evaluation of the SOPA-model is then easily implemented using *sopa=A*X+N* with *N* a similarly generated matrix containing the non-linear terms. Note that also for these simulations, the *sopaparam* parameter set is used. So a direct connection between the behavioural model and the numerical verification is established.

The flexibility of this simulation towards various SOPA-architectures without rewriting the simulator code, is one of the major advantages of the numerical simulation approach. Other important advantages are:

- The numerical simulation is very fast and stable due to the use of the ODE-PACK routines.

- The numerical model is limited to the ideal filter blocks only. No parasitics are involved in the simulation.

- The numerical accuracy can be fully controlled from the octave environment.

- Simulation results are directly available in the memory of the octave environment and can be visualised using the available octave-functions.

The high flexibility enables a fast prototyping environment. However when the simulator reaches a semi-stable form, a faster implementation is mandatory in

Listing 5.5. C++ implementations

```
for (i=0;i<(Internal_sopaparams["filterorder"])(0).scalar_value() ...
    ;i++)
{
    int size = (int)(((Internal_sopaparams["sopaorder"])(0). ...
        scalar_value())+
        (int)(Internal_sopaparams["filterorder"])(0). ...
            scalar_value());
    Matrix Ahulp(size,size,0.0);
    if (filtercorner.is_matrix_type()) {
        if (ffilsm.length()<(Internal_sopaparams["filterorder" ...
            ])(0).scalar_value())
        {
            warning("Filterparameters_do_not_correspond_with_ ...
                filterorder,_taking_first_element_only\n");
            Ahulp(i,i)=-2*PI*ffilsm(0,0);
            if (i>0) Ahulp(i,i-1)=2*PI*ffilsm(0,0);

        }
        else
        {
            Ahulp(i,i)=-2*PI*ffilsm(i,0);
            if (i>0) Ahulp(i,i-1)=2*PI*ffilsm(i,0);
        }
    }
    else
    {
        Ahulp(i,i)=-2*PI*ffils;
        if (i>0) Ahulp(i,i-1)=2*PI*ffils;
    };
}
```

order to do fast design space exploration. Using the liboctave library, a fairly easy conversion towards a C++ implementation is possible.

Listing 5.5 shows the C++ implementation of part of the octave m-file implementation of code snippet 5.4. The programs becomes lengthier due to the necessary extra type checks that need to be implemented. Since this code gets compiled, and thus does not need to be interpreted before execution, the evaluation time will be much smaller. Numerical solutions of differential equations require more function evaluations than solely the time intervals on which an output was requested. Therefor the evaluation speed of this evaluation should be as high as possible.

Note that the parameter structure is in this case implemented as a C++ standard library map. Functions are provided to sync this parameter set to the *sopaparam* global value implementation and the other way around. Since the *Internal_sopaparams* parameter set is not accessible by the interpreter, its values cannot be altered un-intendedly. Therefor it provides a more rigid implementation of the parameter structure concept.

In the ODEPACK simulations, it is also possible to simulate a SOPA with finite comparator gain and finite integrator gain. Other system non-idealities were not implemented. These will be further elaborated in the next section. Resistive coupling of two SOPA amplifiers is also possible to simulate using the ODEPACK-simulator.

2.4.2   Hspice/Eldo Simulations

circuit simulator The ODEPACK numerical simulator is an excellent tool to investigate the system on the architecture and is a fast verification-tool of the implemented behavioural models. Its implementation, however, does not provide an easy interface towards the circuit level. For this an interface from the octave framework with standard circuit level simulators like ELDO and HSPICE. The interface should incorporate the parameter structure concept and the circuit simulators should also be usable at a high level, comparable with the ODEPACK simulators to verify the simulator results between two abstraction levels and to move easily between them during the design.

To implement these abstraction levels, the system is implemented by two sets of files which are part of the model library. At the top level this can be regarded as one main *spice* .cir-file and one *octave* .m-file. The spice file contains the test-bench architecture containing a set of .include control cards which can be controlled from the octave framework. By properly selecting the files which are included every different architecture can be selected and a simulation on the requested abstraction level can be performed. This, of course, assumes that the necessary models, in this case the necessary parameterised sub-circuits, are available in the model library. The top-level octave file provides the interface to the *sopaparam* parameter set. These interfaces contain the necessary instructions to create the right *.include* statements and the instructions to set all parameters of the parameterised building blocks. The design knowledge is also incorporated in these files by adding dependencies between the low level parameters like for instance the resistor and capacitor value of the loop filter and the higher level parameters, like e.g. the corner frequency. In this way the portability of the complete circuit is eased.

The behavioural models are implemented using ideal elements and controlled sources. Figure 5.4 shows some examples of the implemented be-

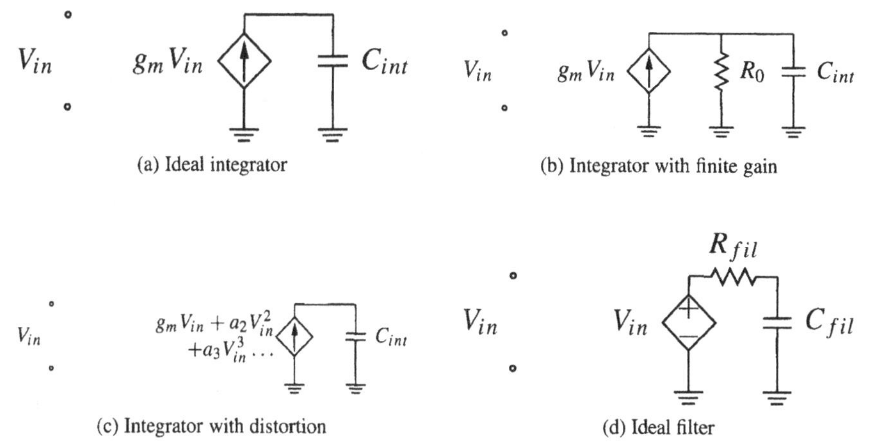
V_{in}
$g_m V_{in}$
C_{int}
(a) Ideal integrator
V_{in}
$g_m V_{in}$
R_0
C_{int}
(b) Integrator with finite gain
V_{in}
$g_m V_{in} + a_2 V_{in}^2 + a_3 V_{in}^3 \ldots$
C_{int}
(c) Integrator with distortion
R_{fil}
V_{in}
V_{in}
C_{fil}
(d) Ideal filter

Figure 5.4: Examples of behavioural building blocks implemented in the spice-like simulators

havioural building blocks. All building blocks have an input with infinite impedance, so the mutual influence is eliminated. The ideal elements need to be equipped with a number of dummy inputs to be compatible with the library model with the highest number of inputs. This is a task for the library-manager and can easily automated. This to be able to switch to lower abstraction levels without extra effort.

The same techniques can be applied to design the building blocks using a top-down approach. The numerical simulators can also be driven through a secure shell tunnel on another workstation as the one the octave framework is running on. In this way the load can be spread over the available computing power.

The biggest disadvantage is that the accuracy settings of the spice-like simulators is a tangle of option control cards and the creation of a specific set of options to reach the desired accuracy tends to be a tedious and time-consuming task. Moreover, since the SOPA, is an asynchronous hard non-linear system, the variable time-step algorithms of spice-like simulators produce errors in the transition points. This gives a sort of jitter problem, which is expressed as a noise contribution between the signal frequency and the limit cycle oscillation. Since the numerical algorithms of spice-like transient simulations are not designed for these type of hard non-linear circuits, the time-step should be fixed to small values in order to simulate the behavioural model at low input amplitudes.

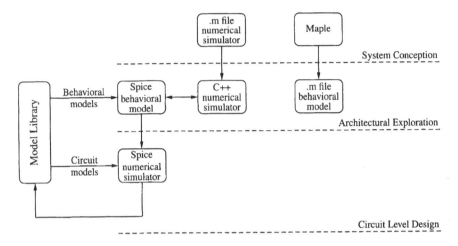

Figure 5.5: Overview of the different numerical simulation techniques and their applicability in the design cycle.

2.4.3   Comparison between numerical methods

Although at the highest level the numerical simulators provide the same results, all methods come into play in different stages of the design. The octave *.m file implementation* is important for the development of new architectures which do not fit into the reference model of figure 4.1. Due to its fast prototyping abilities, this implementation is the one to use in the very first part of the design and for the first design only.

If the models are in a more finished state, the porting to a C++ implementation will enable a fast exploration of the existing techniques. Besides the behavioural models, it should be the main tool in the high level synthesis layer of the design plan (see figure 5.1) as model verification tool. Techniques exist to verify the validity of the describing function behavioural models [Bergen et al., 1982, Blackmore, 1981]. These techniques give a quantitative idea of the filter and amplitude criterion of the describing function analysis. Since these techniques require more computing power and are not correlated with the other methods of the design flow, these techniques were not implemented. During the design, the error bound control is replaced by numerical simulations.

In order to smooth the transition towards the circuit design level, the spice-like simulators should be used at least once at the behavioural level. In this way the implementation of the architecture can be verified and compared with the behavioural model. This is also important to organise the feedback from circuit level restrictions towards the behavioural level.

Figure 5.5 summarises the previous thoughts. The system conception level is the level on which the basic system concepts are analysed or changed. Nor-

Table 5.2: CPU time consumption comparison of the different numerical models

Simulator	Mean	Standard Deviation
.m file simulator	2297 s	8 s
C++ simulator	52 s	0.4 s
ELDO simulator	3648 s	102 s
Behavioural OCTAVE model	45 ms	7 ms
Behavioural MAPLE model	9.7 s	0.4 s

mally one would not have to enter this level for a redesign. Only for drastic architectural changes some work needs to be redone in this level. The main tools are MAPLE for the symbolic analysis part and numerical simulations using an .m-file model to verify the behavioural models. On this level, flexibility of the tool-set is the most important consideration. If the newly chosen set of architectures[3] is stable, the use of MAPLE and the .m-file simulations would be too time consuming. A translation towards C++ and an octave implementation of the results of the symbolic analysis speeds up the architectural exploration. On this level simulation speed and compatibility with the upper and lower levels are the things to keep in mind. The obtained results are translated into a spice behavioural model. This to rule out implementation errors between the architectural model and the actual circuit level implementation. The design is than continued using the spice interface towards a full implementation. Via the model library, design consideration coming from the implementation level can be fed back into the architectural layer.

Table 5.2 compares the execution times between the different software tools. For this test the third order distortion of a third order SOPA is calculated for a signal of 1 MHz and an amplitude of 0.1 V. The limit cycle frequency is set around 17 MHz. The test platform used is a SUN Ultra 60 workstation. The test includes setting one parameter, simulation, visualisation and specification extraction. The results are the calculated mean and standard deviation of 20 consecutive experiments. All models have the same abstraction level.

2.5 Measurement Interfaces

The measurement interfaces can be split into two parts: the signal generating interface and the interfaces to the measurement equipment.

---

[3]In this context, a model like the one of figure 4.2 is considered to be one architecture. A system concept change would be for instance the coupling of 2 SOPAs. The choice of the order of the SOPA lies on the architectural exploration level.

2.5.1   Signal Generation

The signal generation part could also be placed into the model library section, but since its original use was the generation of DMT-signals for the Rohde and Schwartz AMIQ signal generator [Rohde and Schwartz,], it has been catalogued as a measurement interface. Next to this interface, the signal generator provides outputs to the octave and spice numerical simulators. In this way the design can be measured and simulated with the same DMT input. The basic properties of the signal generator are :

- Different band plans are implemented in the generator code : ADSL up- and downstream, symmetrical VDSL and the fibre-to-the local exchange VDSL band plan.

- Each carrier is 2 bit QAM modulated.

- The signal generator delivers signals with a customisable minimal CF.

- A simple pre-warping filter has been implemented to compensate the signal roll-off for high frequency signals in a non zero order SOPA as been explained in section 4.3.4.

- Different output formats are supported : floating point matrices for the octave numerical simulators, binary output for the AMIQ signal generator and a more generic binary output with customisable accuracy. For the spice interface, a sub-circuit-file is generated containing one sine-wave current source for every carrier in the DMT-signal. In this way an ideal time-continuous xDSL signal is available to the spice circuit description.

- The parameter structure approach is also applied in this section, enabling an easy portability between the different design needs.

2.5.2   Equipment interfaces and post-processing

Since the octave environment provides a set of built-in functions to analyse and visualise signals, it is logical to also include an interface to the measurement results and to post-process these results in the same environment. Another big advantage of integrating the measurement set-up in the design flow is the direct feedback towards future designs. Also, this enables the fast detection of measurement or design errors by comparing measurement results with simulation results. Process variations can be extracted and fed back to the simulator in order to find unexpected results.

As an example of the post-processing environment, listing 5.6 shows a snippet of the code to detect the MTPR of an obtained spectrum. Basic image processing techniques are used to determine the gap between the mean power of the tones and the filled antenna-tone. No knowledge of the applied band-plan, nor the settings of the spectrum analyser, like frequency span, video or resolution bandwidth are required to obtain these figures. To extract all those parameters, a 4 step algorithm is followed. In the first step the tone-spacing is calculated by taking an FFT of the obtained spectrum. The peak in the so obtained 'spectrum' will denote the frequency at which peaks occur in the measured spectrum, and is thus related with the tone-spacing. In order to find the off-set frequency at which the tones start, a Dirac-impulse train with the same tone spacing is cross-correlated with the measured spectrum. If the resolution bandwidth is not exactly matched with the tone spacing of the signal, the power of a particular tone may be spread over several bins. To compensate this 'Moiré-effect', the Dirac impulses are each perturbated to find the exact position and energy content of each spectral peak. Listing 5.6 is in this case a good example of the ease of creating post-processing scripts in the octave environment.

Other post-processing tools that were developed include an estimator of the amplitude roll-off at high frequency. Using a limited equalisation filter and the signal generation software, this is a direct feedback to the measurement setup, emulating a digital pre-processor.

3. Conclusions

This chapter has a bridging function between the previous chapter, which is more theoretical, and the chapter which describe the real implementations. By developing a design plan from the mathematical analysis a structured design methodology is developed. The design plan is aimed for minimal design iterations and thus a minimal design-time. Also the insertion of design knowledge in the design flow is eased to provide a faster redesign. The design plan is built up in several layers to structure the design. The layers should be followed successively. To avoid iteration steps, each layer not only provides an optimal design point but also the updated boundaries of the design space.

The choice of a process technology is the most important starting condition to every design. The choices for different CMOS technologies have been explored and the scaling laws have been applied to a standard SOPA design. As an important conclusion, it was stated that if full scaling has been used to go to the next process generation, the performance gain will be less sub-linear. This full scaling property is common for most sub-micron technologies that do not provide output transistors that can operate at higher voltages.

Also, in the selection of a coupling transformer, high voltages are beneficial for the design of a SOPA. The basic limitation from the transformers point-of-view has been described. By taking this basic analysis the output power, output resistance and bandwidth specifications construct the first limitations of the complete design space. The exploration of the available transformers together with the choice of process technology forms the design initialisation layer.

Listing 5.6.   Octave code to extract the MTPR of a measured spectrum

```
[fout ,Ysout ,fmiss ,Ysmiss]=tonefinder(f,Ys)
% Step 1 : Find the spacing between tones
        [spacespec]=fft(repeat(Ys-mean(Ys),1));
        [m,n]=max(spacespec(8:length(Ys)/2));
        n=length(spacespec)/(n+6);

% Step 2 : create tonemask
        tonemask=zeros(1,rows(Ys));
        for k=1:floor(length(Ys)/n) , tonemask( ...
           ceil(k*n))=1; endfor

% Step 3 : basic cross-correlation to determine  ...
    the start position
        Max=tonemask*Ys;
        bestmask=tonemask;
        for k=2:n-1,
                tonemask=shift(tonemask,1);
                m=tonemask*Ys;
                if m>Max,
                          Max=m;
                          bestmask=tonemask;
                endif
        endfor
        tonemask=bestmask;

% Step 4 : Small perturbations
... Rest of the code snipped
endfunction
```

In a next layer, the design space is further reduced and an optimal parameter set is derived for the design. This layer is made up by ordering the behavioural models in a design chain that involves the coupling between the technology and design specifications and the SOPA parameters. The driving force in the last stage of this high level synthesis layer is the estimation of the power efficiency. Therefor a power estimator is developed which uses the technology parameters and the SOPA parameters as inputs. The power estimator does not need pre-knowledge of possible implementations, but can be refined if the design plan is used for a redesign.

To support this particular design methodology a set of CAD-tools has been assembled. Since the SOPA is a hard non-linear system, full numerical simulations are very time-consuming. The use of the derived behavioural models, speeds up the high level synthesis with almost 5 orders of magnitude. However speed is not the only criterion for a design tool. The derived algorithms from the mathematical analysis are translated in a set of simulation tools which are compatible at a certain level, but that can also take the design to a lower abstraction level.

Chapter 6

REALISATIONS IN MAINSTREAM CMOS

THE presented theoretical analysis of chapter 4 and the constructed design plan of chapter 5 needs to be verified by practical implementations. For this two test-chips were constructed in a mainstream 0.35 μm CMOS technology. The supply voltage of this technology is 3.3 V.

The first test-chip is the implementation of a zeroth order SOPA which is the closing piece of a feasibility study for the SOPA concept. It complies to G-Lite specifications.

The second test-chip aims at full ADSL and VDSL compliance to form a multi-standard xDSL line driver for the CO. The economic benefit for a multi-standard xDSL line driver cannot be underestimated. Not only does it save development cost but also installation and maintenance cost will be reduced. Both chips aim at maximum efficiency since power consumption at the CO-side is a major issue as been explained in chapter 2.

For every chip the specific goals will be explained at first and their implications on the design choices are further elaborated. All building blocks will be thoroughly discussed together with their implications on the overall system behaviour. For every test-chip the most important and relevant measurement results are given, and each description is concluded with a critical overview of the most important strong and weak points of the tested chip.

The presented chips will prove that using the SOPA concept line drivers for xDSL can be constructed with superior efficiency. The chips are extensively compared with the present state-of-the-art. The result of this comparison is that the presented implementations not only advance the state-of-the-art but are more than three years ahead of competition.

1.   A Zeroth Order SOPA in .35 μm digital CMOS

1.1   Goal of the Test Chip

This test chip [Piessens and Steyaert, 2001] was designed as the concluding proof of a feasibility study [Piessens and Steyaert, 2002b] to investigate the use of a SOPA for the high demanding xDSL specifications. Therefor it was not directly the intention to fully comply the specifications for a certain member of the DSL family. Another goal was to prove the possibility to implement a switching power amplifier in a mainstream sub-micron CMOS technology. As it has been shown in section 5.1.1.1, a .35 μm CMOS technology should enable the design of an xDSL compatible line driver. Since from .35 μm on, technologies use full scaling, it is suspected that the improvement in performance will be marginal when going to deeper sub-micron if extra technology steps for higher voltage output transistors were left aside.

The goals led to the following design choices:

- Since feasibility of the SOPA technique was the main goal, the implementation was deliberately kept simple. The SOPA is of zeroth order, so no circuit imperfections due to the integrators would be observable.

- No analogue extensions should be used for the design of the chip. This allowed integratability in a true digital CMOS technology. Poly-poly capacitances and high-ohmic poly resistances were not to be used in the design.

- The design is targeted to the least demanding xDSL specification if linearity is concerned, being the G-Lite requirements.

- The limit cycle frequency should not be too high. Parasitic effects due to too high switching signals were unwanted for they would mask the working principles of the SOPA concept. However, the feasibility of the full ADSL bandwidth is explored. Therefor the limit cycle frequency is fixed at 4 times the Nyquist bandwidth of ADSL, being 8.8 MHz.

The block schematic of the realized SOPA is given in figure 6.1. The two basic SOPA building block will be integrated on the same die to provide better coupling. Every block will be integrated, except the line transformer and the coupling capacitance C_{tank}. The used transformer was an Midcom 50702R transformer [Midcom 50702R,]. This transformer is designed to drive ADSL signals on a 100 Ω line with a transformer ratio of 1:2. To emulate a higher transformer ratio the used load resistor will be scaled appropriately.

Since the loop filter will be made without any analogue extension, it is foreseen that it will consume a considerable amount of silicon area. Therefor the filters order is kept minimum, being 3. Therefor the filters cut-off frequency will be $\frac{1}{\sqrt{3}}$8.8 MHz=6.1 MHz.

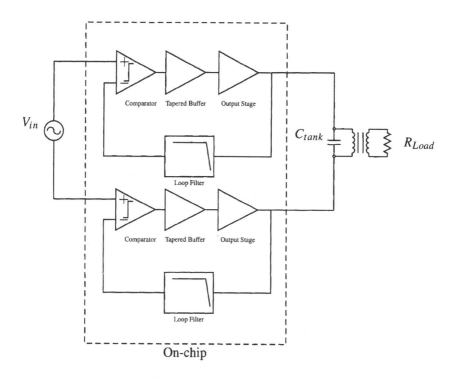

Figure 6.1: Block schematic of the first prototype.

1.2   Building Block Design

1.2.1   The Output Driver

The output driver needs to be designed according the criteria for the design of a regular class D output stage, as presented in section 2.5.3.2. The output resistance is determined by taking the required output power into account. G-Lite requires 16.3 dBm on a 100 Ω load. The rms voltage required to deliver this 43 mW is 2.07 V. If the required CF is set to 15 dB or a factor of 5.6. The peak voltage is thus 11.6 V. since the limit cycle frequency should be higher than this peak-voltage a factor of 3 has been chosen for the maximal voltage range. Since the SOPA system is a bridge type line driver, the total required voltage has to be divided by 2. The voltage one building block has to be able to deal with is thus 17.5 V. To reach this value in a 3.3 V technology, a minimal transformer ratio of 5.3 is necessary. Since integer values are preferred as been demonstrated in section 5.1.1.2, a transformer ratio of 6 has been chosen. This is in accordance with the obtained values from the possible MTPR reach in the presence of thermal noise, as been demonstrated in figure 4.20. This leads to an equivalent load resistance of $100/6^2 = 2.7$ Ω.

Table 6.1: Transistor sizing of the output stage.

	NMOS	PMOS
W	1.000 mm	1.944 mm
L	.35 μm	.35 μm
C_{in}	1.5 pF	3.0 pF

The output resistance determines the output drain efficiency. It is chosen to only have a maximum 10% efficiency decay in the output stage. Since the output resistance is given by :

$$R_{DS} = \frac{1}{KP\left(\frac{W}{L}\right)\left[V_{GS} - V_{T0} - \frac{V_{DS}}{2}\right]} \tag{6.1}$$

With the technological parameters of the Alcatel 0.35 μm technology, this led to W/L-value for the NMOS of 2844. This lead for the output driver to the W and L values given in table 6.1 together with their resulting input capacitance. The output gates are not connected. In this way a steering mechanism can be constructed to avoid shoot-through currents as will be explained in the next paragraph.

1.2.2 The Tapered Buffer

The the (W/L) ratio of the PMOS over the NMOS for a unit inverter was set to 4.3, leading to a sizing of W=1 μm, L=.35 μm for the NMOS and W=4.3 μm, L=.35 μm for the PMOS . The input capacitance of a unit inverter for these values is C_{i0} =8.1 fF. The optimal tapered buffer would thus have

$$n_{opt} = \ln\left(\frac{C_{in}}{C_{i0}}\right) - 1 = 5 \text{ stages.} \tag{6.2}$$

This leads to a scaling factor of 3.3 between every inverter. Using (5.5) and (5.4), this will lead to a buffer delay of 70 ps, which is more than high enough for a 8.8 MHz limit cycle frequency. The choice was made to use this extra delay time, and combine the tapered buffer with a non-overlapping clock circuit. The resulting schematic of the used tapered buffer is shown in figure 6.2. By using the feedback over the nor-gates, the switching signals are shifted with the delay of the inverters in the loop. Since there is still margin in the achievable delay time, the first up-scaled inverter $n = 3.3$ is skipped in order to increase this non-overlapping time.

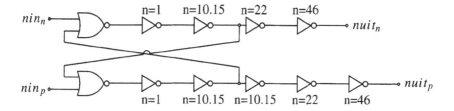

Figure 6.2: Schematic of the used tapered buffer

The total delay of the tapered buffer with k inverters and scaling factors n_i can be approximated by :

$$\tau = t_{d0} + \sum_{i=1}^{k-1} \frac{n_{i+1}^2}{n_i} t_{d0} \tag{6.3}$$

The total delay should be lower than 10% of the limit cycle period. Two other considerations were taken into account to obtain the final sizing of the inverter chain :

1 The power consumption of the inverter chain is proportional with the total capacitance in the chain :

$$P = V_{DD}^2 f \sum_{i=1}^{k} n_i^2 C_{in0} \tag{6.4}$$

By skipping the inverter with the highest scaling factor $n_i = 112$, a power saving of 7.3 mW can be obtained.

2 Since the line is coupled with a transformer, a fast off switching of the current can create voltage spikes at the output, which can be lead to destructive oxide breakdown. In the SOPA structure this problem is worthwhile to consider since the output is fed back and thus returns on an input gate.

These consideration led to the sizing of the inverter chain as depicted in figure 6.2.

1.2.3   The Comparator

The comparator needs to be continuous time, since no clocking is allowed in the SOPA-loop. The consequence of this is that the comparator cannot be put into an meta-stable 'reset' state, and no offset cancellation techniques can be used [Allstot, 1982]. Figure 6.3 shows the schematic of the used comparator. It is a three stage circuit consisting of

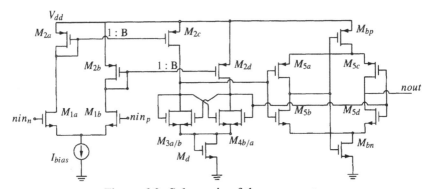

Figure 6.3: Schematic of the comparator.

1 A pre-amplifier that functions as a voltage to current converter. The pre-amplifier increases the comparators sensitivity a bit and insulates the inputs from switching noise coming from the positive feedback circuit [Baker et al., 1998].

2 A decision circuit that utilises positive feedback for a fast comparison.

3 A post-amplifier that transforms the output voltage swing of the decision circuit into digital signal levels.

The Pre-amplifier.   The pre-amplifier converts the input voltage into an input current for the decision circuit. The $(V_{gs} - V_T)$ of the input stage needs to be chosen as small as possible, since it will directly set the switching point voltages as will be explained in the next paragraph. Therefor the $(V_{gs} - V_T)_1 = 0.2$ is fixed.

Since offset cannot be compensated, the minimisation of mismatch of the input pair is mandatory. For a minimal mismatch in mirrored current, the overdrive voltage of the current mirror transistors M_{2abcd} has been chosen to be higher, namely $(V_{gs} - V_T)_2 = 0.5$.

Another source of mismatch is the mismatch of the input stage formed by transistors M_{1ab}. Therefor the lengths of the used transistors are chosen to be non-minimal. The gain bandwidth of the pre-amplifier is given by :

$$GBW = \frac{I_{bias}}{2\pi \left((1 + B)C_{ox}L_2 W_2\right)(V_{GS} - V_T)} \tag{6.5}$$

The input stage is deliberately made very fast since the total comparatordelay through the comparator-tapered buffer chain is the limiting factor for the speed of the SOPA system. The GBW of the input stage is set to 125 MHz. The lengths of both transistor pairs are chosen equal. The current relation couples

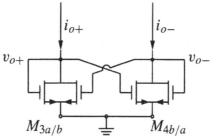

Figure 6.4: Close-up of the decision circuit

in this way the widths of both transistor pairs :

$$W_2 = W_1 \frac{(V_{gs} - V_T)_1^2 \, K P_n}{(V_{gs} - V_T)_2^2 \, K P_p} \tag{6.6}$$

By filling in, (6.6) in (6.5), the maximal allowable transistor length $L = L_1 = L_2$ can be calculated so the pre-amplifiers time constant is lower than 10% of the limit cycle period :

$$L = \sqrt{\frac{2(V_{GS} - V_T)_2^2 K P_p}{2\pi(1 + B)C_{ox}GBW(V_{GS} - V_T)_1}} \tag{6.7}$$

For a B-factor of 1, this results in a transistor length of 5 μm. The required bias current will be determined by the requirements of the decision circuit. When the bias current is fixed, all transistor sizing of the comparatorpre-amplifier can be calculated.

The Decision Circuit. The comparatordecision circuit is formed by transistors M_{3ab} and M_{4ab}. Transistor M_d is inserted to set the output DC voltage to a more appropriate voltage for the following stage. The a-transistors are transistors coupled in a diode configuration, the b-transistors are cross-coupled to form positive feedback. For symmetry reasons, $\beta_{3a} = \beta_{4a} = \beta_a$ and $\beta_{3b} = \beta_{4b} = \beta_b$. The heart of the decision circuit is depicted in figure 6.4.

To study the working principle of the decision circuit of figure 6.4, a large signal study is mandatory. When i_{o+} is much larger than i_{o-}, transistors M_{3a} and M_{4b} are on, while M_{3b} and M_{4a} are off. Under these circumstances v_{o-} is approximately 0 V. The voltage v_{o+} can thus be calculated as :

$$v_{o+} = \sqrt{\frac{2i_{o+}}{\beta_a}} + V_T \tag{6.8}$$

If now current i_{o-} is increased and i_{o+} is decreased, the decision takes place when v_{o-} equals the threshold voltage of M_{4b} and this transistor enters the

Table 6.2: Transistor sizes of the pre-amplifier and the decision circuit

	M_1	M_2	M_3	M_d
W	60 μm	24 μm	8 μm	5.6 μm
L	5 μm	5 μm	5 μm	5 μm

saturation region. The current through transistor M_{4b} equals :

$$i_{o-} = \frac{\beta_b}{2}(v_{o+} - V_T)^2 = \frac{\beta_b}{\beta_a}i_{o+} \qquad (6.9)$$

The complementary transition yields a switching point :

$$i_{o+} = \frac{\beta_b}{\beta_a}i_{o-} \qquad (6.10)$$

Relating these equation to the current relation of the input pair, the switching point voltages (V_{SPH}, V_{SPL}) can be calculated as :

$$V_{SPH} = \frac{(V_{GS} - V_T)_1}{2}\frac{\frac{\beta_b}{\beta_a} - 1}{\frac{\beta_b}{\beta_a} + 1} \qquad (6.11)$$

and, by the symmetry properties :

$$V_{SPH} = -V_{SPL} \qquad (6.12)$$

The transfer function for half a decision circuit is :

$$\frac{v_{o+}}{v_{in}} = \frac{g_{m1}B}{g_{ma} - g_{mb} + 3g_o + sC_{o+}} \qquad (6.13)$$

in which C_{o+} denotes the capacitance at the output node and g_o the output conductance of the transistors M_{3ab} and M_{2c} who are taken equal for simplicity. Note that a maximum gain is reached as $g_{ma} = g_{mb}$ which corresponds with a comparator on the edge of hysteresis. Note that B was taken 1 for mismatch and speed reasons, so the necessary bias current can be calculated (6.13) if the same condition on the gain bandwidth is applied to the complete system ($GBW > 10\omega_{LC}$). The necessary bias current is 100 μA. From this value all transistor sizings can be calculated. Table 6.2 summarises all transistor dimensions.

The Post-amplifier. The chosen post-amplifier is a self-biased differential amplifier [Baze, 1991]. The amplifier works for applications where the common-mode range is relatively limited. Due to the stacked diodes in the decision circuit this is certainly the case. The big advantage of this differential amplifier is that it is self-biased through negative feedback, therefor the amplifier is

Table 6.3: Transistor sizes of the post-amplifier circuit

	M_{5bd}	M_{5ac}	M_{bp}	M_{bn}
W	4.95 µm	20 µm	5 µm	20 µm
L	5 µm	4 µm	5 µm	4 µm

- less sensitive to variations in processing, temperature and supply

- capable of supplying switching currents that are significantly greater than the quiescent current.

The structure of the post-amplifier can be seen at the right side of figure 6.3. Transistors M_{bp} and M_{bn} form the biasing circuit and operate in their linear region. Therefor the source voltages of transistors M_{5abcd} can be put very close to the supply rails. Therefor the working principle of the M_{5abcd} transistors can be regarded as a regular CMOS inverter. The gain can be calculated to be :

$$A \simeq \frac{g_{m5bd} + g_{m5ac}}{g_o} \tag{6.14}$$

A gain of 40 dB could be met. Table 6.3 summarises the dimensions of the post-amplifier.

1.2.4 The Loop Filter

The loop filter is a simple RC chain configuration. The values for the resistors and capacitors are equal over the filter. This causes three closely located poles instead of one pole with multiplicity 3. From a working principle point-of-view this does not change a lot. The limit cycle frequency will decrease a bit, but since the phase transition is more smoothly, it will be less sensitive to value variations.

To determine the sizes, the noise density of this filter needs to be taken into account. Since a telephone line has a background white noise of -140 dBm, the noise in the feedback filter should not exceed this level. The total resistance is thus the result of the following calculation.

$$\overline{v_R^2} = (4kTR) \times \text{filter order} \times \text{transfo-ratio} \tag{6.15}$$

This results in a maximal resistance value of 10.06 kΩ which leads to resistance of 3.35 kΩ and a capacitance value of 9.3 pF. Since the linearity of these elements will determine partly the overall systems linearity and to make it implementable in a digital CMOS technology a metal realisation of both the resistors and capacitors is mandatory. Table 6.4 lists the most important process parameters to construct resistors and capacitors in a digital technology. It

Table 6.4: Most important parameters to create resistors and capacitors in a digital CMOS technology

Parameter	Value
C_{M1-M2}	0.0384 fF/μm^2
C_{M2-M3}	0.0384 fF/μm^2
C_{M3-M4}	0.0384 fF/μm^2
C_{M4-M5}	0.0384 fF/μm^2
ρ_{M1}	55 mΩ/$\square$
ρ_{poly}	2 Ω/$\square$
ρ_{n-well}	1 kΩ/$\square$

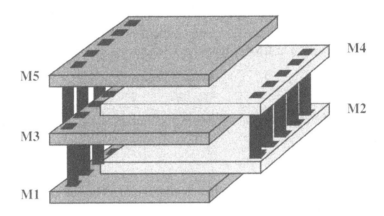

Figure 6.5: 3D representation of a metal-metal capacitor

becomes clear that the devices will be huge in silicon area. If all 5 metal layers are combined in a wafer structure [Aparicio and Hajimiri, 2002] as demonstrated in figure 6.5 the total capacitance per μm^2 is only 0.1536 fF/μm^2, which is almost a factor of 10 lower than the values obtainable with a similar technology including analogue extensions. These poly-poly capacitances have values up to 1.1 fF/μm^2.

To integrate the 9.3 pF capacitance, an area of 60500 μm^2 is required. For this implementation, this led to rectangle of approximately 420 μm$^2 \times$ 144 μm^2. For the integration of the resistances only metal 1 has been used. To obtain the total resistance of 3.35 kΩ, 61000 squares are mandatory. This means a total line length of more than 42.6 mm for a line with minimal length of 0.7 μm. Rolled together in a serpentine, the resistor occupies an area of 450 μm $\times$ 122 μm which is as big as the capacitance area.

1.2.5 The Complete Schematic

The complete schematic is shown in figure 6.6. Note that two of those are implemented on the same die and are externally connected by means of a line transformer.

1.3 Layout Considerations

1.3.1 Chip photograph

The resulting chip photograph is depicted in figure 6.7. The area consumption of the comparator is clearly negligible with the area consumption of the RC-filter and the output driver. The total chip area is 4.6 mm^2.

A ring of decoupling capacitances is laid out around the chip. These capacitances are MOS-capacitors for their high value per area. For the SOPA is a switching amplifier it is very important to decouple the supply voltage of the analogue blocks as much as possible.

Simulations have revealed the maximal allowable bonding wire inductance that is allowed for not degrading the performance. Since the chip was designed for classical wire bonding, this meant reducing the inductance by placing at least 8 bonding wires in parallel for ground and supply of the output driver.

1.3.2 Electro-migration

The area of the output driver is mostly determined by the electro-migration rules of the technology. Since the output current equals :

$$I_{peak} = \text{transfo-ratio} \times \text{CF} \times \sqrt{\frac{P_{out}}{R_{Load}}} \qquad (6.16)$$

For the presented SOPA amplifier these peak currents are up to 0.5 A. To prevent electro-migration for these high currents, the width of the conductors that carry these currents needs to obey the following limitation :

$$W = K\, I_{eq} f(T) + \Delta W \qquad (6.17)$$

In which I_{eq} denotes the equivalent rms current in a period of 200 ns, ΔW the tolerances on the width and the technological constant K can be calculated as :

$$K = \frac{1}{T_{Al}\, SC_{Al}\, I_{max}} \qquad (6.18)$$

In this formula T_{Al} denotes the Aluminium thickness and SC_{Al} the step coverage of the Aluminium conductor. I_{max} is the maximal current through a Aluminium conductor with a cross section of 1 μm^2. To avoid electro-migration $I_{max} = 2$ mA/μm^2. $f(T)$ denotes the temperature dependence of the electro-

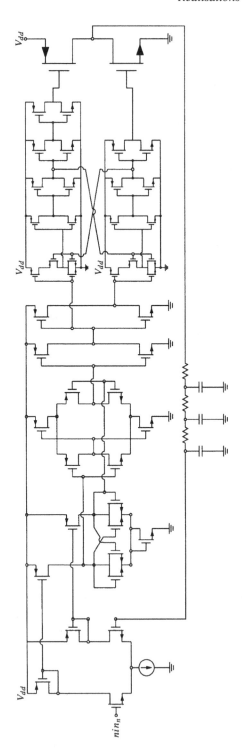

Figure 6.6: Complete schematic of the implemented zeroth order SOPA.

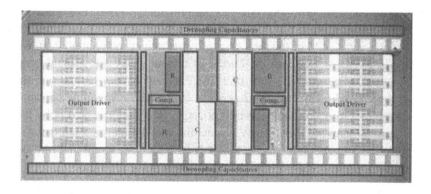

Figure 6.7: Chip photograph of the implemented zeroth order SOPA

Table 6.5: Technological metal electro-migration rules

		M1	M1,2,3	M5
T_{Al}	[nm]	396	495	765
SC_{Al}	[ ]	0.95	0.95	0.95
K	[μm/mA]	1.33	1.06	0.68
ΔW	[μm]	0.05	0.07	0.08

migration phenomenon and is given by :

$$f(T) = \exp\left(9.8\left(1 - \frac{415}{T}\right)\right) \qquad (6.19)$$

Table 6.5 lists the most important technological parameters. The use of metal 5 is advisable. Due to the fact that the metal 5 layer is much thicker, it can carry almost double the current of a metal 1 conductor of equal size. To reach the higher metal layers sufficient vias need to be used. The electro-migration rule for minimal number of vias N_{min} is given by :

$$N_{min} = K_C \, I_{eq} \, f'(T) \qquad (6.20)$$

In this formula K_C is a technological constant which is 1/0.9 mA for contacts and 1/1.88 mA for a regular via. The temperature dependence is slightly changed to :

$$f'(T) = \exp\left(11.7\left(1 - \frac{415}{T}\right)\right) \qquad (6.21)$$

In order to easily take these consideration into account while making the layout, the following strategy has been followed :

1 The output stage is divided into 128 parallel output stages. One basic building block is laid out in such a way that all conductors have a sufficient width to cope with 1/128 of the total current. The high current supply conductors are laid out parallel with the border of the sub-cell. The ground conductor uses M1, to provide substrate shielding, while V_{DD} is laid out on top of the ground conductor. This provides a parasitic decoupling capacitance distributed over the complete output stage. The output signal is by means of a maximum number of vias, directed to metal 4 and 5. Both metals will form the output conductor. In this way its width can be limited. All this is shown in figure 6.8(a) where the basic driver sub-cell is shown.

2 Four basic inverters are then placed together to form 1/32 of the total output driver. Since the supply conductors are laid out on the side, by a simple mirror action of the basic block, the supply conductor widths are doubled between two 1/128 driver stage. Another similar mirror operation forms the complete 1/32 block as depicted in figure 6.8(b). The only extra action is the connection of the output conductors with a metal4-metal5 stack to form an appropriate conductor.

3 This mirror and copy operation is repeated 3 more times to form a combination of 8 1/32 inverter blocks. This is depicted in the layout of figure 6.8(c) A large enough conductor to carry the required current for this building block is laid out along the perimeter of this building block. Since the interior conductors are sufficiently large, this ring conductor is actually over-dimensioned for normal operation with an equal current spread. From a layout point-of-view, the initial state is reached again, with a partial driver that has sufficiently large supply conductors running at its perimeter.

4 The final output stage can thus be constructed by a mirror and copy operation. However to provide a low inductance, multiple bonding wires are necessary and the choice was made to connect every 1/4 building block with its own set of bonding pads. Since the interior pads are all connected with upper pads to form the output nodes, these interior pads do not compromise wire bonding, for these bonding wires my touch each other. The layout of the final output stage is depicted in figure 6.8(d).

1.3.3 Other Considerations

As can be clearly seen on the chip photograph of figure 6.7, both output drivers are placed directly opposed to each other. In this way the thermal gradients are supposed to run as almost straight lines along the length of the chip. By this and the symmetrical layout strategy, the sensitive analogue inputs of the comparators are in this way located at the same thermal gradient, providing better matching.

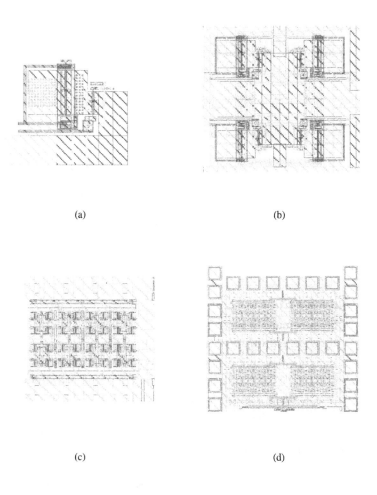

(a)   (b)

(c)   (d)

Figure 6.8: Various stages in the layout of the output driver.

By construction the driving nodes of the output driver are laid out as a binary tree. For simultaneous switching the tapered buffer is laid out inter-digitised. The tapered buffer is split into several parallel tapered buffers. The tapered buffer driving the NMOS output transistor and the one steering the PMOS are laid out alternating. In this way the complete output stage should switch approximately at the same time.

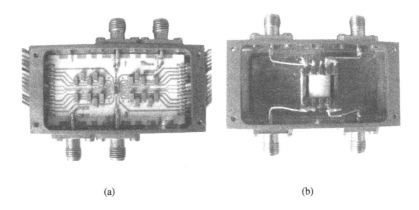

(a)   (b)

Figure 6.9: Housed chip under test and Midcom transformer

1.4   Measurements

1.4.1   Measurement Set-up

The chips were wire bonded on a ceramic thick film substrate. The sub-strate was mounted in a copper beryllium box for better shielding. An example is shown in figure 6.9(a). All supply lines were on the substrate decoupled by 470 nF capacitors. The big drawback of the use of these substrates was the rather high resistivity of the substrate. Since the equivalent load at the output is mere 2.4 Ω and the output paths had resistances near 0.7 Ω, much efficiency degradation was due to this parasitic resistance. In the efficiency measurements, these losses were calibrated out. Also the used transformer and the used equivalent line impedances were mounted, so all could be connected using SMA cable for better shielding, as shown in figure 6.9(b).

The test signals were generated using a Rohde and Schwartz AMIQ signal generator as described in chapter 5. Time domain measurements were performed with the Tektronix 7854, while for the spectral measurements a HP 3585B spectrum analyser was used. The bias currents was set by an off-chip resistor which is decoupled by a parallel connection of several off-chip capac-itors, all mounted on a wire board.

1.4.2   Sine wave inputs

At first the response to a sinusoidal input signal has been measured. Fig-ure 6.10 shows the output spectrum up to 20 MHz of the amplifier for an input signal of 200 kHz is applied. A Spurious Free Dynamic Range (SFDR) of 56.4 dB has been obtained, without any extra filtering, except the combination

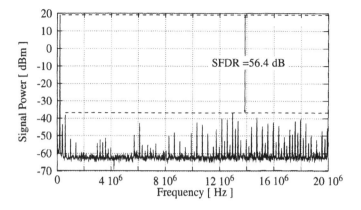

Figure 6.10: Measured spectrum up to 20 MHz

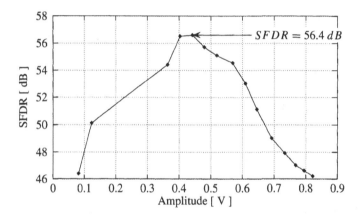

Figure 6.11: Measured SFDR versus input amplitude

of the line transformer and the coupling capacitance C_{tank}. The equivalent load resistance is 2.4 Ω.

The maximum dynamic range for a sinusoidal input signal with an input frequency of 200 kHz is plotted in figure 6.11. A maximum SFDR of 56.4 dB is reached for an input amplitude of 0.4 V. Above this amplitude, the distortion components become dominant. This was expected, for only a zeroth order SOPA has been used. This figure hardly changes for higher signal frequencies. For a 800 kHz input signal an SFDR of 54.4 dB has been measured. For a 900 kHz sine wave the SFDR has decreased to a level of 51.2 dB. For higher signal frequencies the limit cycle gets to much attracted by the signal, so the linearising effect drops drastically.

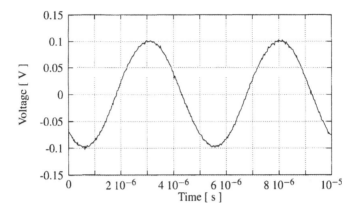

Figure 6.12: Measured sine wave signal at the line input.

The corresponding measured oscilloscope signal is shown in figure 6.12. The waveform is measured at the secondary of the line transformer, thus on the equivalent line impedance. As can be clearly observed from figure 6.12, the limit cycle oscillation is almost completely suppressed. This has been illustrated further in the measurements of figure 6.13. For this two measurements are super-imposed on each other. The spectrum on the load resistance is measured and compared with the according spectrum on one of the primary inputs of the line transformer, this is thus directly at the output of one of the two SOPA amplifiers. One can clearly observe the suppression of the limit cycle frequency and the even harmonics around the limit cycle frequency. Figure 6.13 shows a good accordance with the theoretical results [Piessens and Steyaert, 2003b] of figure 4.24. The measured limit cycle frequency is located at 3.8 MHz, this is almost a factor 2 lower than the value for which the SOPA was designed. This is probably due to mismatch on the metal-metal capacitors and resistors. This explains the bandwidth limitations to 800 kHz.

Another thing one can learn from figure 6.13 is that from the output of one SOPA amplifier towards the line impedance the noise floor decreases with almost 20 dB. This points out that the noise is substrate noise, due to the switching behaviour of the SOPA. Since the mean switching frequencies are common mode towards the output, the switching noise at one primary is supposed to correlate for a large part with the noise on the other primary. If no input signal is applied, this becomes more pronounced. The switching is in that case completely synchronised and the measured output noise drops again with almost 20 dB extra. Since the used .35 μm technology has a low ohmic substrate, the guard rings do not influence the substrate noise.

Figure 6.14 shows the measured efficiency of the SOPA amplifier for a sine input of 200 kHz. It is compared with the values of other state-of-the-art line

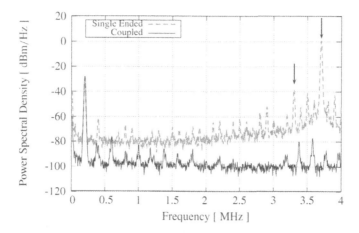

Figure 6.13: Illustration of the limit cycle suppression by oscillator pulling.

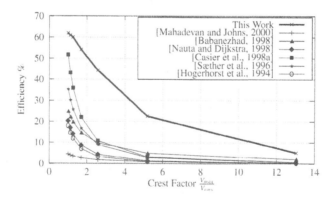

Figure 6.14: Plot of the efficiency versus crest factor, and compared with other state of the art line drivers

drivers at the time of publication of the SOPA. The SOPA has a superior efficiency due to its switching nature. More important is the fact that this efficiency does drop slower than inverse proportional with the crest factor. The theoretical efficiency should be more than 90%. Further simulations with improved technology models revealed that this lower efficiency is due to the fact that the widths of the output transistors are chosen too narrow.

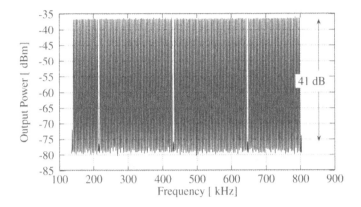

Figure 6.15: Measured output spectrum of the zeroth order SOPA when applying a DMT - signal with a CF of 5.6

1.4.3   xDSL-performance

As been shown in section 2.3.3.1 the SFDR, does not characterise the capabilities of a line driver to be usable for ADSL. The better performance measure for xDSL performance is the MTPR. This figure is measured by applying a DMT-modulated signal to the line driver with several tones missing. Figure 6.15 shows an output spectrum when 185 tones are applied to the line driver. Tones 1-32 are missing, to simulate a downstream signal. Tones 50, 100 and 150 are missing to act as antenna tones. The CF of the signal is 15 dB. An MTPR of 41 dB was achieved.

1.4.4   Overview

Table 6.6 gives a summary of the most important measurement results and compares them with the G-Lite specifications.

1.5   Discussion of the Results

1.5.1   Comparison with the state-of-the-art

Table 6.8 shows the performance of the most state-of-the-art published Class AB power amplifiers at the time of introduction of this test chip. The efficiency of the presented SOPA is almost twice as good as the maximum efficiency of the presented works. The distortion is better than all of them except for [Sæther et al., 1996] and [Casier et al., 1998a] but

- [Sæther et al., 1996] operates at 10 V supply with a load of $1k\Omega$

- The bandwidth of [Casier et al., 1998a] is an order of magnitude smaller

Table 6.6: Performance Summary

Parameter	Measured	ADSL-Lite
Technology	0.35 μm CMOS	
Supply Voltage	3.3 V	
Output Power	18.7 dBm	16.3 dBm
Voltage Gain	0.9	
-3 dB Bandwidth	800 kHz	500 kHz
Mean switching frequency	3.8 MHz	
Maximum efficiency	61%	
	48% incl. meas. set-up	
SFDR @ f_{sig} = 200 kHz	56.4 dB	
MTPR	41 dB	34 dB

- The bandwidths of publications [Mahadevan and Johns, 2000, Nauta and Dijkstra, 1998, Babanezhad, 1998] are the -3 dB bandwidths. They however cannot guaranty the SFDR to stay in this -3 dB border. For the SOPA however the SFDR doesn't drop significantly in its bandwidth.

Table 6.7 shows some relevant published switching type audio power amplifiers. The Total Harmonic Distortion (THD) is in the same order of magnitude. Since these power amplifiers are for audio applications, their bandwidth is about 40 times smaller.

- The audio power amplifier reported in [van der Zee and van Tuijl, 1998] it consists of a non-integrated class AB power amplifier in parallel with a switching device. It also uses a BCD-process with Bipolar, CMOS and DMOS devices, so it can operate at a 36 V supply.

- [Philips et al., 1999] describes an audio power DAC where the modulation is done in advance by a clocked digital noise-shaper. The discrimination of this clock frequency and the up-converted noise is not reported.

The bandwidth of the presented SOPA is larger than all of them. This is due to the difference in mean switching frequency over bandwidth ratio (4.75 for the SOPA).

These tables prove that this SOPA outperforms the state-of-the-art line drivers.

1.5.2 Strong Points

The most remarkable results of this test-chip included :

- The feasibility of a switching line driver for xDSL has been proven in practice. The specification for G-Lite were met.

Table 6.7: Comparison with published switching type power amplifiers

Ref.	Technology	Swing V_{pp}	BW MHz	THD dB	efficiency %	Note
[Mahadevan and Johns, 2000]	3.3V/.35 μm CMOS	2	160	-50.8...-42	4	THD published for a band of 20 MHz
[Nauta and Dijkstra, 1998]	2.4V/.35 μm CMOS	1.4	130	-50	20	
[Sæther et al., 1996]	10V/3 μm CMOS	7	5.5	-77	35	$R_L = 1k\Omega$
[Casier et al., 1998a]	3.3V/0.5 μm CMOS	5	0.060	-68	51	
[Babanezhad, 1998]	3.3 V/0.4 μm CMOS	5.78	107	-45	24.7	THD = -40 dB at 100MHz, optimised transformer

Table 6.8: Comparison with published Class AB Line Drivers

Ref.	Technology	Swing V_{pp}	BW MHz	$\frac{f_{switch}}{BW}$	THD dB	efficiency %	Note
[Philips et al., 1999]	5/0.5 μm CMOS	4	0.02	50	-60	70	modulation done by clocked noise shaper
[van der Zee and van Tuijl, 1998]	36/BCD	31	0.02	27.5	-60...-90	85	Not fully integrated

- Due to its switching nature, a superior efficiency compared with even the present state-of-the-art was reached.

- The measured results confirmed the results of the modelling effort. The working principles presented in chapter 4 were shown in practice.

- The oscillator pulling into synchronisation could be confirmed by measurements.

- It has been proven that it is possible to construct the line driver in digital sub-micron CMOS technology. This opens other possibilities to integrate high efficiency power amplifiers in a mainstream technology.

- The feasibility to use metal-metal wafer capacitances and metal resistors for the construction of linear loop filters has been shown.

- The chips has proven to be reliable and no electro-migration effects were noted during measurements.

1.5.3   Possible Improvements

Some remarks, however, need to be made :

- The limit cycle frequency was almost a factor of 2 too low. This is possibly due to an incorrect modelling of the metal-metal capacitors.

- The linearity is too low for full ADSL. this was to be expected from the behavioural modelling. A higher order SOPA is mandatory to reach more demanding specifications.

- Substrate noise is a major noise contributor. Much care needs to be taken to lower this noise contribution by further decoupling of the analogue part. Due to the oscillator pulling it could be shown that substrate noise was the major contributor.

2.   A Third Order SOPA in .35 µm CMOS
2.1   Goal of the Test Chip

Since the feasibility study proved the SOPA concept, the goals for this next test chip needed to be more ambitious. The primary goal of the research project in total was the design of a line driver for ADSL [Piessens and Steyaert, 2002a]. To deal with the limited linearity of the zeroth order design, integrators need to be included. Higher order SOPAs have an MTPR that decreases with increasing frequency as has been calculated in section 4.3.4. Since the linearity specifications for the higher frequent VDSL system are less stringent as that of an ADSL modem, this opens up a way to create a multi-mode line driver. This is depicted in figure 6.16. The theoretical MTPR curves for a zeroth order and

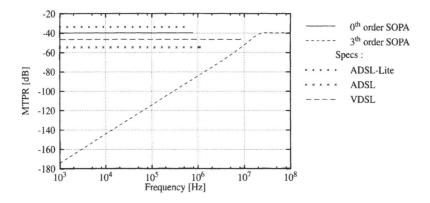

Figure 6.16: Possible MTPR reach versus frequency, compared with several xDSL specifications

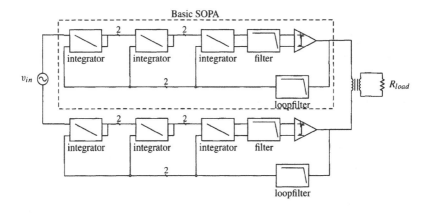

Figure 6.17: Block schematic of the implemented third order SOPA

a third order SOPA are plotted next to a rudimentary spectral mask of several xDSL-flavours.

The choice for a third order SOPA is derived from figure 6.16 and is a trade-off between the extra power consumption of an extra integrator and the dynamic power-consumption of a higher switching frequency. The limit cycle frequency has been chosen to be 20 MHz, for it seemed the limit of the technology. The complete architecture can be seen in figure 6.17.

The first prototype was clearly limited by switching noise. The integrators form an extra 'antenna' for substrate noise, so it is tried to minimise these effects by making all analogue building blocks fully differential. In figure 6.17, this is explicitly denoted by marking the signal paths as 2 wire busses. As a

Table 6.9: Transistor sizing of the output stage.

	NMOS	PMOS
W	1.000 mm	2.944 mm
L	.35 μm	.35 μm
C_{in}	1.5 pF	4.5 pF

consequence the integrators need to be realised as four input analogue, continuous time, integrator circuits.

The loop filter has been split in a second order section following the last integrator and a first order part in the feedback path. In this way the forward filter diminishes high frequent noise.

The choice has been made to reuse the layout techniques of the output driver from the zeroth order line driver. The technology thus stays the .35 μm CMOS technology, but this time the analogue extensions were used to create high ohmic resistances and poly-poly capacitors. This to reduce the chip area.

The limit cycle frequency will be pushed to the limits of technology, since the VDSL downstream bandwidth is 8.5 MHz. The used Midcom transformer, however hasn't got sufficient bandwidth. Therefor the ADTT1-6 [ADTT1-6,] has been chosen for this design.

2.2 Building Block Design

2.2.1 The Output Driver

This third order line driver should be able to transmit the full ADSL output power, for a multi-mode xDSL driver is the goal of this test-chip. As such, the design considerations of the output stage are the same as those for the zeroth order design. The zeroth order design, however, showed an efficiency decrease due to a insufficient dimensioned PMOS driver stage. A better modelling led to an increase of the PMOS width by a factor of 1.5. The resulting sizes are summarised in table 6.9.

2.2.2 The Tapered Buffer

In the zeroth order design, the tapered buffer started from a minimal sized inverter. This however is a a start from a too small amplifier, since the sizes of the comparator are already larger. In the third order design this inconsequence has been annihilated by resizing the minimal inverter to W=2.5 μm, L=.35 μm for the NMOS and W=7.36 μm, L=.35 μm for the PMOS . The input capacitance of a unit inverter for these values is C_{i0} =15 fF. The optimal tapered

Table 6.10: Scaling factors along the tapered buffer

	n_1	n_2	n_3	n_4	n_5	n_6
Scaling factors	5.5	13	30.7	72.2	170	261

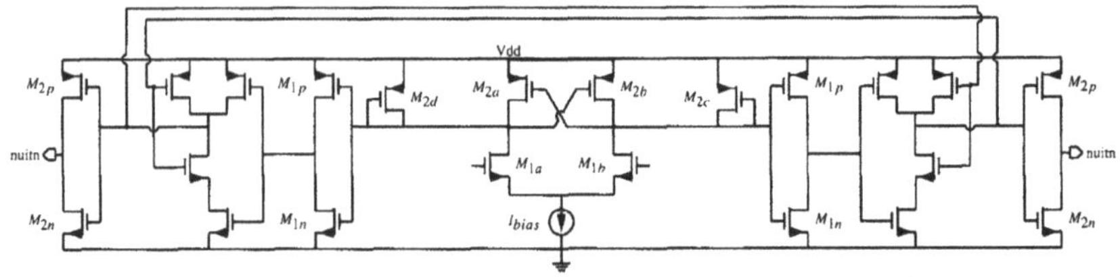

Figure 6.18: Used comparator in the third order SOPA

buffer would thus have

$$n_{opt} = \ln\left(\frac{C_{in}}{C_{i0}}\right) - 1 = 6 \text{ stages.} \tag{6.22}$$

This leads to a scaling factor of 2.7 between every inverter. Using (5.5) and (5.4), this will lead to a buffer delay of over 25 ns which becomes very close to the actual speed limitation. The implemented scaling factors are given in table 6.10. Or the comparator should be made very fast, or the limit cycle frequency will be determined by the total delay in the tapered buffer comparator block.

Due to this timing constraint the no-through-current circuitry is omitted in this design. This is solved by putting an extra delay at the end of the NMOS tapered buffer with scaling factor 261. This delay enforced the output stage transistors never to be on at the same time.

2.2.3   The Comparator

The comparator used in this design is depicted in figure 6.18. The structure of the zeroth order comparator needed to be changed since :

- The speed of the comparator should follow the increase in limit cycle frequency. Therefor a faster implementation had to be chosen.

- On the other hand, the comparator should consume less power. Since the line driver is aimed for multi-mode xDSL, its power consumption should take the lower output power of VDSL into account, so the total efficiency for VDSL is not too low. This consideration becomes even more eminent since the integrators also take a huge amount of the power budget.

As a result, the decision circuit and pre-amplification circuit are merged into each other to save bias current. This is possible by the insertion of part of

Table 6.11: Sizings of the comparator for the third order SOPA

	M_{1ab}	M_{2ab}	M_{2cd}	M_{1n}	M_{1p}
W [μm]	7	1.2	12	4	3.87
L [μm]	.35	.7	7	1	3

	M_{2n}	M_{2p}	M_{nandn}	M_{nandp}
W [μm]	2.55	7.36	2	2.87
L [μm]	.35	.35	.35	.35

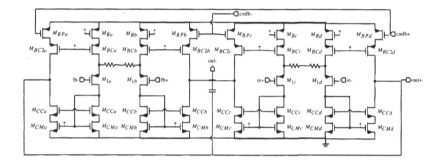

Figure 6.19: Schematic of the continuous time integrator

the loop filter in front of the comparator. Possible kick-back noise from the switching decision circuit are thus filtered out, by the decoupling of the integrator output and the comparators input.

The post-amplifier has been exchanged for a nand set-reset latch. This extra positive feedback should enable faster comparison. The outputs are by the latch faster restored to full digital levels.

The bias current was set to 18 μA. All dimensions are given in table 6.11.

2.2.4 The Integrators

The integrators in general are by far the most demanding building block in the design of the third order SOPA. The first integrator will set the overall performance and will thus be the most demanding. Since all other integrators' non-idealities are shaped by the preceding building blocks, their specifications will be less demanding. The integrators have been integrated as $g_m - C$ filters. For reasons of project timing however, the g_m-stages are the same for every integrator, and the different coefficients have been realised by down-scaling the integration capacitance.

Table 6.12: Transistor dimensions for the integrated g_m stage

	M_{1abcd}	M_{CMabcd}	M_{CCabcd}	M_{Babcd}
W [μm]	37	9.6	9.6	37
L [μm]	.35	.35	.35	.35

	M_{BCabcd}	M_{BPabcd}	$M_{BC2abcd}$
W [μm]	37	37	37
L [μm]	.35	.35	.35

Figure 6.19 depicts the schematic of the continuous time integrators. Two degenerated g_m-stages are coupled together to form a 4-input integrator. The currents are mirrored on the integration capacitance by means of a high swing mirror-stage with mirror-factor $B = 1$. The integrators are made fully differential to provide better supply noise immunity. Degeneration is necessary to meet the high linearity specification for an xDSL system. Equation (6.23) gives the third harmonic distortion for the degenerated g_m-stage.

$$HD_3 = \frac{-1}{32} \frac{g_m^2}{I_{Bias}^2 (g_m R_E + 1)^3} V_{rms}^2 \qquad (6.23)$$

For this implementation, a degeneration factor $g_m R_E = 5.1$ sufficed. Optimal system performance has been found when the unity gain frequencies are doubled from the systems input towards the comparator input. The implemented values are 0.5 MHz, 1 MHz and 2 MHz respective. The unit gain frequency of the integrator is given by :

$$GBW = \frac{g_m}{2\pi (g_m R_E + 1) C_{int}} \qquad (6.24)$$

Also the noise of the first integrator will be fully visible at the output. Since the noise will be dominated by the degeneration resistor, its value can be 10 kΩ to generate -132 dBm noise at the output. From (6.23), it can be calculated that at least a degeneration gain of 2.5 is necessary to meet a -74 dB distortion specification. This sets the required g_m to 150 μS and thus the necessary bias current to 15 μA for every dual-input stage. The necessary integration capacitances will thus be 20 pF for the first integrator and 10 pF and 5 pF for the following stages. All these design parameters lead to the transistor dimensions summarised in table 6.12.

2.2.5 The Loop Filter

From the discussion of the comparator, it becomes clear that the delay through the comparator - tapered buffer chain limits the speed of the complete SOPA-system. In principle, no loop filter is mandatory to set the limit cycle frequency, since it will only limit the bandwidth. The loop filter however is kept in the system to lower the signal swing at the integrator's input and to reduce kick-back noise. The structure is the same R-C stage as in the zeroth order design, but is made up by poly-poly capacitors and high-ohmic resistors. The resistance value is set to 10 kΩ and the capacitance value is 50 fF. The latter is chosen a bit too small, it can be enlarged to more than 200 fF without jeopardising the system's behaviour.

2.3 Layout Considerations

The chip photograph of the processed third order SOPA is depicted in figure 6.20. The total area consumption is 6.76 mm^2. All described building blocks are clearly visible.

Most layout considerations are similar to the ones of the zeroth order design and are described in section 6.1.3. Some more things however are still worth mentioning :

- The bigger inverters of the tapered buffer were not left out in this design for delay reasons. These larger inverters are also laid out inter-digitised and can be clearly observed.

- The integrators occupy a large area. This is mostly due to resistors which have been laid out over a large area to lower their mismatch.

- Also the up-scaling of the integration capacitors can be clearly seen.

- To cope with the parasitic resistance of the thick film substrate used in the measurements and the related problems as been described in section 6.1.4, the outputs are put at the same side of the die. This enabled also a more 'square' layout.

- Substrate and supply noise was a major issue in the previous design. To lower their contributions many decoupling capacitances are put on the die as close as possible to every analogue building block. In this way, every building block has its own local decoupling.

2.4 Measurements

2.4.1 Measurement Set-up

The chip is wire bonded on a ceramic substrate and then mounted in a copper beryllium box. The output paths were made as wide as possible to minimise the parasitic resistance. Also the supply and ground conductors were

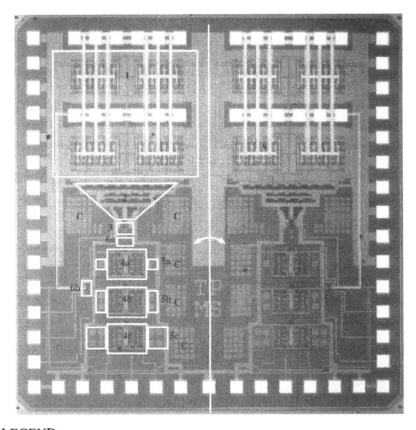

LEGEND :

1:Output Driver, 2: Tapered Buffer, 3: Comparator, 4abc: g_m-stages,

5abc: Integration capacitance, 6a: Loop filter in forward path, 6b: Feedback loop filter,

C: Decoupling capacitances

Figure 6.20: Chip photograph of the third order SOPA

widened compared with the previous design and more decoupling was put on the substrate. The bias currents were set by potentiometers on a wire board and sufficiently decoupled as close as possible to the chip connectors. Everything is fed from a single supply as can be seen in figure 6.21.

The measurement equipment used was the same as for the zeroth order design except for the mounted transformer. Due to the high bandwidth specifications for VDSL the RF ADTT1-6 transformer is used. Two transformers are coupled in such a way that they provide a 1:4 transformer ratio. The output impedance is a 25 Ω resistance, so the equivalent load impedance is 1.5625 Ω. No extra tank capacitance was used to ease oscillator pulling.

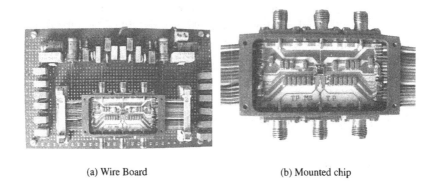

(a) Wire Board   (b) Mounted chip

Figure 6.21: Measurement set-up for the third order SOPA

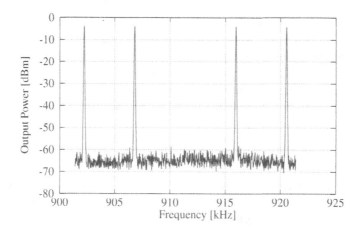

Figure 6.22: A 56 dB MTPR line measurement around tone 226

2.4.2   ADSL Characterisation

An MTPR measurement has been performed to derive the ADSL specifications. For these measurements a DMT signal consisting of 256 tones with a tone-spacing of 4.3125 kHz is applied to the line driver. Tones 1-32 are left blank to form the upstream band and Tones 50, 100, 150, 200 and 226 are left out as antenna-tones. Figure 6.22 shows a 20 kHz zoomed spectrum around the antenna-tone at the highest and most critical frequency.

An MTPR of 56 dB has been measured for an output power of 100 mW. The total current consumption is 64 mA out a 3.3 V supply. This gives a 47 %

power efficiency for driving a 100 mW ADSL signal with a crest factor >5. For a 134 mW output power an MTPR of 56 dB is still achieved. In this case the total current consumption is 76.6 mA from a 3.3 V supply, meaning a 53 % power efficiency. A higher output power was not feasible due to the resistance of the output driver stage. This output power is not sufficient for resistive back termination, but provides more than enough headroom for active back termination.

The standby power, when no signal is applied is around 0.1 mA. This proves the power savings due to the oscillator synchronisation.

2.4.3   VDSL Measurements

Also for VDSL measurements a test signal with a downstream Cabinet Deployment power spectral density mask (FTTCab) [Wang, 2001] and a crest factor of >5 has been applied to the line driver with the same test set-up as for the other measurements. The tone spacing is 4.3125 kHz. The tones in the downstream band between 1.622 MHz and 3.75 MHz and the tones between 5.2 MHz and 8.5 MHz were activated. The other tones are left unused as antenna-tones in the upstream bands. The noise floor is measured to be −103 dBm/Hz. The precautions that were taken during the design to lower substrate noise were thus effective. No spurious tones were observed at the antenna-tones even for double the necessary output power. The measured power efficiency for this signal is 20%.

Figure 6.23 shows a measurement of the output spectrum zoomed in on the first antenna-tones within the transition bands together with some tones at the beginning and end of the pass bands. Since these antenna-tones are closest to the pass-bands taking a 175 kHz transition region into account, figure 6.23 shows the worst case linearity measurement for a VDSL-signal.

2.4.4   FTTEx Deployment

To prove the full potential of using a SOPA line driver at the Central Office, th FTTEx deployment scenario mask [Wang, 2001] has been used, since it offers the worst case power spectrum to be delivered by the line driver. For this test the ADSL and VDSL downstream bands are filled with tones. The other bands are kept as antenna tones. The PSD measured at the line is −40 dBm/Hz for the ADSL-band (142 kHz - 1.1 MHz) and −60 dBm/Hz in the VDSL downstream-bands (1.622 MHz - 3.750 MHz and 5.2 MHz - 8.5 MHz). The tone-spacing is 4.3125 kHz and the crest factor of the signal is set to be >5. The output power measured at the line with a resolution bandwidth of 30 kHz is given in figure 6.24.

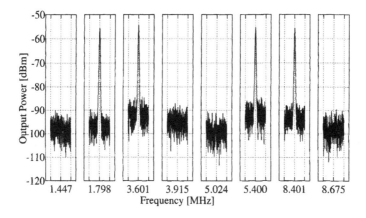

Figure 6.23: Measurement of the critical VDSL spectrum tones, measured at the line

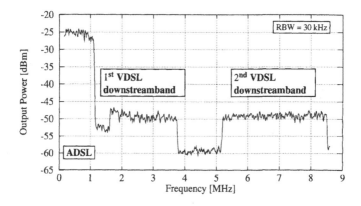

Figure 6.24: Measured output power spectrum on the line when ADSL and VDSL signals are applied together

2.4.5   Summary

Table 6.13 summarises the most important measurement results and compares them with the related xDSL specifications.

2.5   Discussion of the Results

2.5.1   Comparison with the present state-of-the-art

Table 6.14 summarises the state-of-the-art up to now. The presented material is more as double as efficient as the most efficient solution up to now. If the year of publication is regarded, it becomes clear that the presented test-chips advance the state-of-the-art and are more than 3 years ahead of all competition.

Table 6.13: Performance summary

Parameter	Measured	xDSL specs
Technology	0.35 μm CMOS	
Supply Voltage	3.3 V	
Bandwidth	> 8.6 MHz	8.5 MHz (VDSL DS)
MTPR	56 dB	55 dB
Output Power ADSL	−38.7 dBm/Hz	−40 dBm/Hz (DS)
Out of Band PSD	<−103 dBm/Hz	−100 dBm/Hz
Crest Factor	> 5	>5
Efficiency	47 %	

There is still no indication that the presented specifications will be beaten. Note also that all other chip-sets use higher voltage techniques. As been presented in chapter 4, a SOPA-design will only benefit from going to higher supply voltages.

Also, it needs to be noted that all solutions use a DMT-based modulation scheme for reasons of the high CF involved. The reported efficiencies for the HDSL and symmetrical G.SHDSL solution are measured for a signal with a CF of 3.2 instead of the 5.5 for other DMT-based chip sets.

2.5.2   Strong Points

After evaluation of the third order SOPA test-chip, it has to be concluded that :

- The SOPA concept works and it is usable even for high-end applications like xDSL.

- It is possible to create a highly efficient line driver for DMT modulated signal form a single supply.

- Due to oscillator synchronisation the standby power consumption is very low. This is another important advantage of the proposed structure.

- Even in mainstream CMOS, the bandwidth of VDSL can met with an in-band linearity sufficient for ADSL. This enables the use of this line driver as a real CO multi-standard DSL solution.

- The effects of substrate and supply noise were decreased by differential design and thorough decoupling. The presented techniques can also be useful for other analogue design in sub-micron CMOS technologies with low ohmic substrates.

Table 6.14: Comparison with state-of-the-art published and commercial solutions

Reference	xDSL-specification	Power Consumption	Supply voltage	Efficiency
This work:				
Zeroth order (2001)	G-Lite	164 mW	3.3 V	61%
Third order (2002)	ADSL/VDSL	211 mW/141 mW	3.3 V	47%/20%
[Zojer et al., 2000]	G-Lite	2.38 W	60 – 150 V	1.8%
[Kappes, 2000]	HDSL	>155 mW	3 V	<14%
[Laaser et al., 2001]	G.SHDSL	307 mW	5 V	7.3%
[Benton et al., 2001]	G-Lite and full ADSL	1.9 W / 2.65 W	48 V	5.3%
[Cresi et al., 2001]	ADSL	1.1 W	15 V	9%
[Pierdomenico et al., 2002]	ADSL	744 mW	6 V	13.4%
[Sabouti and Shariatdoust, 2002]	ADSL	740 mW	24 V	13.5%
[Maclean et al., 2003]	ADSL	590 mW		17%
[Bicakci et al., 2003]	ADSL	700 mW	2.5 – 5 V	14.3%
[Moyal et al., 2003]	VDSL	700 mW	5 V	14.3%
THS6012	ADSL	2.9 W	33 V	3.5%
THS6022	ADSL/HDSL/VDSL	1.65 W	33 V	6%
THS6032	ADSL	1.35 W	33 V	7.4%
AD8393	ADSL	600 mW	12 V	16.7%

- The results from the behavioural modelling have been proven by measurements and their results can be used for further development of self-oscillating line drivers.

2.5.3   Possible Improvements

A next SOPA could gain by taking the following into account:

- The noise is still an issue. Several possible improvements can lower the noise level even further :

  - The choice of sub-micron CMOS is interesting from a research point-of-view, but for real product development a higher voltage technology for the line driver stage is advisable. A higher output voltage in the output stage will decrease the transformer ratio and will lower the output noise, since thermal noise will not be up-transformed with the transformer ratio.

  - A high-ohmic substrate or further development of techniques to lower the effect of switching noise will improve the performance of the SOPA system.

  - The used values for the resistance and capacitance of the used loop filters are actually based on a design error. A lower resistance value and a higher capacitance will lower the total thermal noise of the complete system. With the present values the cut-off frequency of the loop filter is much too high to be of any use in the system.

- In the present design the limit cycle frequency is fully determined by the delay through the comparator-tapered buffer chain. A faster comparator should take the bandwidth of the system higher. In this way it should be possible to span the complete future VDSL bandwidth up to 30 MHz[1].

3.   Conclusions

To prove the theoretical possibilities of the SOPA system, a real implementation is mandatory. Two test-chips were made in a mainstream .35 μm technology.

A first test-chip was designed as the end of a feasibility study for the self-oscillating concept. The chip confirmed the results from the behavioural modelling. The linearising effect of the limit cycle and the oscillator pulling into synchronisation could be clearly observed. The test-chip had a signal bandwidth of 800 kHz and an MTPR of 41 dB. This for a mere switching frequency

---

[1]The VDSL-system specifies a possible bandwidth up to 30 MHz. However up to now only systems with a 8.5 MHz downstream bandwidth were conceived.

of 3.8 MHz. The limit cycle suppression enabled a SFDR of over 56 dB for spurious measured up to 20 MHz. The total efficiency is measured to be 61% for 18.7 dBm. This first test-chip complied with the G-Lite specifications and was the first line driver to comply a DSL spec with such a high efficiency.

Since the first test-chip proved the feasibility, the second test-chip aimed at full DSL compliance. The chip aimed to have a linearity that complied full ADSL standards, which is the most stringent in the xDSL family and the highest bandwidth at the CO-side, being the one of the VDSL standard. In this way a multi-standard line driver is developed that can be installed at the CO and can deliver different standards to the customer. This provides the operator with a lower installation and maintenance cost. By the insertion of continuous time integrators in the forward path, the SOPA went to higher order. Since the first integrator will determine the overall performance, resistive degeneration is mandatory. The measurements showed a full compliance. A 8.6 MHz bandwidth is measured and a MTPR of over 56 dB was reached for the ADSL band. A VDSL downstream measurement was applied to the power amplifier and complete compliance with the standardised spectral mask was observed. The total efficiency is 47%.

In this way it is proven that it is possible to create high efficiency line drivers for xDSL even in a sub-micron CMOS technology. The measurements confirmed the results from the behavioural analysis. Both chips are state-of-the-art and they outperform all other state-of-the-art line drivers up to now in terms of efficiency with a factor of three.

Chapter 7

CONCLUSIONS

IN the introduction, it was stated that there were three distinct motivations for the research activities presented here. The success of this work should thus be measured by the way these three objectives were met. In this chapter, the overall conclusions will be drawn. Every objective will be confronted with its results and valuated.

1. The Objectives

1.1 To build a highly efficient line driver for ADSL

The advent of better digital and DSP technologies opened the evolution to more complicated modulation schemes to cope with lossy channels. These technologies should have undergone a gradual evolution if there wasn't the boom in the communication market by the advent of the Internet. A huge market for broadband communication opened and a huge investment has been put in the development of xDSL technologies. The result was a bandwidth increase of three decades in only one decade of fundamental research and product development. Chapter 1 sketches the growing market for ADSL and puts a finger on the major bottleneck of the development of complete xDSL systems. The line drivers take almost 80% of the power budget. The thermal limitation of a line card to comply with the NEBS norms, limit the number of addressable lines to 24 for present state-of-the-art power amplifiers.

Why the xDSL-family renders present line drivers into power dissipators is a direct consequence of the advanced modulation techniques used. To better understand the reasons for using these techniques the history of twisted pair communication has been described in chapter 2. The telephone has been the most important communication medium for the last century. More than 700 million households are connected to this immense network. It may thus not surprise us that telecom companies want to reuse the copper wires that have been put in the ground during these decades. To fully utilise the Shannon bandwidth of the lossy twisted pair, DMT-modulation is used. This modulation scheme divides his bandwidth in discrete 4.3125 kHz wide channels. Each channel is bit-loaded proportional to the measured SNR in that frequency bin. The big drawback of this modulation technique is the high CF of the resulting time domain signals. It is shown by stochastic analysis in chapter 2 that to have a sufficient low BER, the CF should be higher than 5. Since the efficiency of an ideal class AB line driver is inverse proportional with the CF, a maximum efficiency of 13% can be reached. Several other architectures are summarised in chapter 2 to elevate this problem. It has been shown that none of them really mean a drastic increase in power efficiency. In this work a novel architecture, the SOPA, has been proposed.

Chapter 4 covers this novel architecture and creates a complete behavioural model. The SOPA is a hard non-linear system that has the peculiar property that it exhibits a limit cycle oscillation. This self-oscillation can be used as a natural dither to linearise the input-output relation of a switching power amplifier, which the SOPA still is. An other important feature is the fact that no circuit element in the SOPA system is clocked. By this, the parasitic effect of oscillator pulling into synchronisation is used to reduce the output filter requirements with a factor of 2 in bandwidth and more than 30 dB in suppression. Moreover this phenomenon enables very a low standby power consumption. It has been shown that by the dithering effect a sufficient linearity can be reached to allow G-Lite specifications. If the higher demanding ADSL specifications are required, a higher order implementation needs to be used. A higher order SOPA uses integrators in the forward path. This implements a noise-shaping behaviour. By going to higher order not only the linearity for ADSL can be reached but also multiple standards can be addressed without too much extra power consumption.

The behavioural models are proven by the implementation in silicon. Two test chips have been realised in a mainstream 0.35 μm CMOS technology [Piessens and Steyaert, 2003a]. The complete design of these chips are described in chapter 6. Table 7.1 summarises the most important performance parameters of the implemented line drivers, together with the related xDSL specifications. It proves that the zeroth order SOPA complies with the G-Lite standard, while the third order solution is a true multi-standard line driver that can drive signals with the linearity of an ADSL system and the bandwidth of a VDSL CO driver. Moreover both chips have power efficiencies around 50% for true ADSL signals with a CF higher than 5.

To compare this with the state-of-the-art figure 1.3 from chapter 1 is repeated here. All recently published line drivers from the leading ISSCC are put in this efficiency versus year of publication graph. The evolution of the effi-

Table 7.1: Performance Summary of the two implemented SOPA line drivers

Parameter	Zeroth order SOPA	Third order SOPA	xDSL specification
Technology	.35 μm CMOS 1P5M	.35 μm CMOS 2P5M	
Supply Voltage	3.3 V	3.3 V	
Area	4.6 mm^2	6.76 mm^2	
Bandwidth	800 kHz	8.6 MHz	500 kHz (G-Lite) 1.1 MHz (ADSL) 8.5 MHz (VDSL)
Mean Switch. freq.	3.8 MHz	19 MHz	
MTPR	41 dB	56 dB	34 dB (G-Lite) 55 dB (ADSL)
Output Power	19.9 dBm	21.1 dBm	16.3 dBm(G-Lite) 20 dBm (ADSL)
Efficiency	61%	47%	
Crest Factor	>5	>5	>5

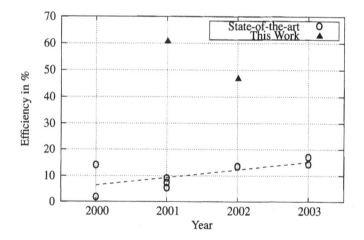

Figure 7.1: Comparison between the power efficiency evolution in present state-of-the-art line drivers and the presented work.

ciency towards higher regions is shown by drawing a dashed trend-line through the other state-of-the-art designs.

1.2   This work advances the state-of-the-art of CMOS power module integration

In Section 5.1.1.1, a scaling law for the implementation of SOPA amplifiers in CMOS has been derived. Starting from the general CMOS scaling laws, it could be derived that the performance of a SOPA amplifier improves with the square of the scaling factor until the scaling laws need to scale the voltage as well. So, when going do deep sub-micron technologies, the performance of a SOPA improves sub-linearly with the scaling factor of the technologies. The only way to break this degradation is to implement (relatively) high-voltage techniques in sub-micron CMOS technologies or to implement drain-source engineering. The bottom line of this all is : the supply should not decrease.

Of course, the SOPA itself is a perfect example of a power amplifier in mainstream CMOS. Much care has been taken to guarantee the reliability of the circuit and a modular layout technique to prevent electro-migration has been presented in section 6.1.3.2. The zeroth order SOPA design also proved that the SOPA system is implementable without using analogue extensions to the process-technology. This has been shown in section 6.1.2.4 by realising the loop filter by metal resistors and metal-metal wafer capacitors.

The SOPA amplifier can also be used in the design of RF-power amplifiers that use envelope extraction and restoration techniques to form a highly efficient power device [Su and McFarland, 1998]. The SOPA is a perfect building block for these kind of applications as a base-band driver.

As a conclusion, it can be stated that the presented techniques broadened the knowledge and the state-of-the-art in the field of CMOS power amplifiers not only for being a state-of-the-art example itself, but by the various techniques presented throughout this book that aid future designs.

1.3   This work advances the knowledge of non-linear analogue design

Chapter 3 introduced the basic mathematical techniques to describe a non-linear feedback system with a low-pass character. Since almost 90% of the analogue systems fall in this category, the obtained results can also be used in other fields.

The presented techniques are applied to the SOPA in chapter 4. Since the SOPA is a hard non-linear system, its output spectrum consists of many distortion and inter-modulation products. The goal was not only to be able to explain the different spectral peaks qualitatively but also quantitatively. In chapter 4 formulas were derived that were able to predict every spectral peak with a very high accuracy as compared with numerical systems. The obtained models are direct formulas, so no iteration is required to obtain the results. Every obtained model has been extensively compared with results form numerical simulation to prove the accuracy. Throughout chapter 4, these comparisons are extensively shown.

The derived models were implemented in the OCTAVE framework to ease the design cycle. This has been described in chapter 5. It was shown that the use of these models reduced the behavioural system evaluation from 3648 s for a high-level spice model and 52 s for a dedicated C++ numerical model to an evaluation time of mere 45 ms. This heavy reduction in simulation speed, allows a flexible architecture exploration and an optimal design even if design centring is required. The increase in speed, however, is not the biggest advantage of the derived models. Most of the obtained models are easily interpretable closed formulas that give a lot insight in the design of non-linear systems.

The SOPA uses oscillator pulling into synchronisation to enable a low standby power and to enable a serious decrease of output filter specifications with a factor 2 in bandwidth and more than 30 dB in out-of-band suppression. Since the coupling of the SOPA is formed by the output load, which is hard to control for a telephone line at the limit cycle frequency, much care has been taken to derive models for this oscillator pulling. The results obtained here can be easily extended for other (parasitic) oscillator attraction phenomena.

Glossary

Abbreviations

ADC	Analog-to-Digital Converter
ADSL	Asymmetric Digital Subscriber Loop
AFE	Analogue Front-End
AM	Amplitude modulation
ASIC	Application Specific Integrated Circuit
AWG	American Wire Gauge, a measure for a cables diameter.
BER	Bit-Error Rate
BRI	Basic Rate ISDN
BiCMOS	Bipolar assisted CMOS transistor/technology
CAD	Computer Aided Design
CF	Crest Factor : this is the ratio between the peak voltage and the rms voltage of a signal
CMOS	Complementary Metal-Oxide-Semiconductor transistor/technology
CODEC	Coder-Decoder
CO	Central Office The switching office of the local telephone company analogue voice signal into a digital bit stream.

CPE	Customer Premises Equipment, the installation at the customers side.
DC	Direct Current. Mostly used as (very) low frequent.
DF	Describing Function If solely the term DF is used, the single sinusoid describing function is meant.
DIDF	Dual Input Describing Function
DMOS	Diffusion Metal-Oxide-Semiconductor transistor/technology
DMT	Discrete Multi-Tone modulation
DSL	Digital Subscriber Line
DSP	Digital Signal Processing
DS	downstream
ECMG	Excess Common Mode Gain denotes the ratio of the gain of a common mode disturbance over the gain for a counter mode disturbance in a coupled SOPA system, expressed in dB. If positive, the common mode oscillation is the one to occur in a physical system.
EC	Echo Cancellation
FEXT	Far-End Crosstalk
FFT	Fast Fourier Transform
FSK	Frequency Shift Keying
FTTCab	Fibre To The Cabinet
FTTEx	Fibre To The Exchange
G-Lite	less performing ADSL-Lite.
GBW	Gain Bandwidth
HDSL	High-speed Digital Subscriber Line
HDTV	High Definition Television
HFC	Hybrid Fibre/Coax
IC	Integrated Circuits

IFFT	Inverse Fast Fourier Transform
ISDN	Integrated Service Digital Network
ISSCC	International Solid-State Circuits Conference
LT	Line Termination
MBD	Missing Band Depth
MTPR	Missing Tone Power Ratio, the ratio between the energy-level in a DMT antenna tone and the output power level.
NEBS	Network Exploitation Board Specifications
NEXT	Near-End Crosstalk
NT	Network Termination
ONU	Optical Network Unit
OSR	Over Switching Ratio, the ratio of the mean switching frequency and the bandwidth of the signal
PAM	Pulse Amplitude Modulation
PAR	Peak-to-Average-Ratio
PCM	Pulse Code Modulation.
PDM	Pulse Density Modulation
PLL	Phase-locked Loop
POTS	Plain Old Telephone Service, the traditional telephony network made of twisted pair wires.
PSD	Power Spectral Density
PSK	Phase Shift Keying
PSRR	Power Supply Rejection Ratio
PSTN	Public Switched Telephone Network
PWM	Pulse Width Modulation
QAM	Quadrature Amplitude Modulated signal
RF	Radio-frequency

rms	root mean square
SFDR	Spurious Free Dynamic Range
SNR	Signal-to-Noise Ratio
SOPA	Self Oscillating Power Amplifier
TCM	Trellis Coded Modulation
THD	Total Harmonic Distortion
TSIDF	Two Sinusoid Describing Function. The TSIDF denotes the gain of a non-linear element for a sinusoidal signal in the presence of another sinusoid with another frequency
US	upstream
VDSL	Very high-speed Digital Subscriber Loop
VGA	Variable Gain Amplifier
xDSL	Digital Subscriber Loop. The term xDSL denotes the whole family of digital subscriber loop technologies. In this thesis we will focus ADSL and VDSL.

Symbols

$\mathbf{0}$	the zero matrix
$_2F_1(a, b, ; c; z)$	the 2-1 hyper-geometric function in the variable z with factors (a, b) and (c)
$\|x\|$	The absolute value of a number x
α	Coupling factor between two coupled self oscillating power amplifiers
$f\|_y^x$	The evaluation of a function f in operating point x and operating point y generate the same result
$f\|_{x,y}$	Evaluate a function f in a operating point (x,y)
$f\|_x$	Evaluate a function f in a operating point x
$\sim$	is proportional to
$\simeq$	is almost equal to
BW	Bandwidth

α_0	The resistive coupling factor between two SOPA amplifiers
A_0	The DC gain of an amplifier
A	The limit cycle amplitude taken at the input of the nonlinearity
A_c	The comparator gain
j	Complex unit $\sqrt{-1}$, see also I
k	The Boltzmann constant $1.3807e - 23$ J/K
L	length of a MOS transistor
M	Magnitude of a complex number $c = M \exp(j\phi)$
$(a)_n$	The Pochhammer symbol, a notation for $\Gamma(x+n)/\Gamma(x)$
arcsin	The inverse sine function.
arctan	The inverse tangent function
RX	Receive path signal
T	absolute temperature
TX	Transmit path signal
W	width of a MOS transistor
BW	bandwidth of a system
C_{in0}	Input capacitance of a unit inverter
C_{int}	integrator capacitance
C_{ox}	The oxide capacitance of a MOSFET
C_{TP}	Channel capacity of a twisted pair
ϵ	total power efficiency of a power amplifier. It is defined as the output power divided by the total power drawn from the supply.
$\mathbf{f}(.)$	denotes the non-linear transfer to the derivative of the state-variables
ϕ	Phase of a complex number $c = M \exp(j\phi)$
f_{LC}	The limit cycle frequency

$\mathbf{g}(.)$	denotes the non-linear transfer to the output of the system
$\Gamma(x)$	The gamma function.
$\overline{\gamma}$	The complex propagation constant in the transmission line model
g_m	transistors transconductance
$HD3$	third order distortion
I	$\sqrt{-1}$
I_{DS}	the drain-source current
$\mathcal{I}m(z)$	The imaginary part of the complex number z
$J_n(x)$	The Bessel function of the first kind and order n
K_P	CMOS transconductance parameter
KP_n, KP_p	Mosfet current factor for the NMOS, resp. the PMOS
$L_f(s)$	Transfer function of a linear loop filter
L_{min}	Minimal gate length of a specified CMOS technology
$\log_2$	the logarithm with base two, also called binary logarithm
μ	magnetic permeability of a material
$N_A(A, B)$	The dual input describing function for a nonlinearity with 2 sinusoidal inputs having amplitude A and B, describing the gain of the signal with amplitude A
$N_A(s, A)$	The single sinusoidal input describing function
n	The order of the SOPA's loop filter
$O(\phi)$	The Landau symbol also called big-O, which denotes that there exists a positive value A so that if $f = O(\phi)$ $|f| < A\phi$
Φ	Denotes a two-port model representation
P_{out}	Output power
P_x	Power consumption of component x
$\mathcal{R}e(z)$	The real part of the complex number z

ρ	resistivity of a material
ρ_x	resistivity of material x
R_L	The load resistance
R_{line}	line resistance
R_{on}	On resistance of a switch
r_{out}	Output resistance of a non-ideal output buffer
R_p	parasitic resistance
SC_{Al}	Step coverage of the Aluminium conductor
$\overline{\sigma_n^2}$	Noise density
s	Laplace variable $= I2\pi f$
T_{Al}	Aluminium thickness
τ	The Greek letter τ denotes the time constant of an exponential settling function
t_{d0}	Delay time of a unit inverter
t_{d0}	Gate delay of a basic inverter
T	Absolute temperature in degree Kelvin
V	A vector containing the state-space variables
V_{DD}	The supply voltage
V_T	MOS threshold voltage
ω	the pulsation in rad/s
X	A boldface uppercase letter denotes a matrix quantity
x	A boldface lowercase letter denotes a vector quantity
$\dot{x}$	A superimposed dot is used as a synonym for the the first derivative with respect to time $\frac{\delta}{\delta t}$
Z_0	Characteristic line impedance
Z_{in}	Impedance seen at the input of the line.
Z_L	Load impedance

Appendix A
Stability Analysis of the Coupled SOPA

1. Stability Criterion

The possible limit cycles are solutions of the Barkhausen criterion, which is a complex equation. Written in polar coordinates this gives :

$$TF = M(A, \omega) \exp(j\phi(A, \omega)) = 1 \qquad (A.1)$$

If a small perturbation is applied to a given solution $\{A_0, \omega_0\}$ of (A.1), the perturbated solutions can be given as :

$$A^* = A_0 + \Delta A \qquad (A.2)$$

$$\omega^* = \omega_0 + \Delta\omega + j\Delta\sigma \qquad (A.3)$$

The perturbation in the rate of change of amplitude has been associated with the frequency term, a device which becomes clear upon thinking of the limit cycle in the form $A_0 \exp(j\omega_0 t)$. This form is the base formulation of a limit cycle if the filter hypothesis holds [Gelb and Vander Velde, 1968]. For the limit cycle to be stable, the small perturbated system solution, being the Barkhausen criterion (A.1), evaluated in A^* and ω^* from (A.2) and (A.3) must also hold :

$$M(A_0 + \Delta A, \omega_0 + \Delta\omega + j\Delta\sigma) \exp(j\phi(A_0 + \Delta A, \omega_0 + \Delta\omega + j\Delta\sigma)) = 1 \qquad (A.4)$$

By definition ΔA, $\Delta\omega$ and $\Delta\sigma$ are small quantities. The Taylor expansion of (A.4) around the equilibrium point, valid to first order terms, after removal of the quiescent terms, becomes :

$$\left(\left.\frac{\partial M}{\partial A}\right|_{A_0,\omega_0} \Delta A + \left.\frac{\partial M}{\partial \omega}\right|_{A_0,\omega_0} \Delta\omega + \left.\frac{\partial M}{\partial \omega}\right|_{A_0,\omega_0} j\Delta\sigma \right) \exp(j\sigma)$$
$$+ jM \exp(j\sigma) \left(\left.\frac{\partial \phi}{\partial A}\right|_{A_0,\omega_0} \Delta A + \left.\frac{\partial \phi}{\partial \omega}\right|_{A_0,\omega_0} \Delta\omega + \left.\frac{\partial \phi}{\partial \omega}\right|_{A_0,\omega_0} j\Delta\sigma \right) = 0 \qquad (A.5)$$

If this complex equation is splitted in its real and imaginary equations, the following system is derived :

$$\left.\frac{\partial M}{\partial A}\right|_{A_0,\omega_0} \Delta A + \left.\frac{\partial M}{\partial \omega}\right|_{A_0,\omega_0} \Delta\omega - M \left.\frac{\partial \phi}{\partial \omega}\right|_{A_0,\omega_0} j\Delta\sigma = 0 \qquad (A.6)$$

$$M \left.\frac{\partial \phi}{\partial A}\right|_{A_0,\omega_0} \Delta A + M \left.\frac{\partial \phi}{\partial \omega}\right|_{A_0,\omega_0} \Delta\omega + \left.\frac{\partial M}{\partial \omega}\right|_{A_0,\omega_0} j\Delta\sigma = 0 \qquad (A.7)$$

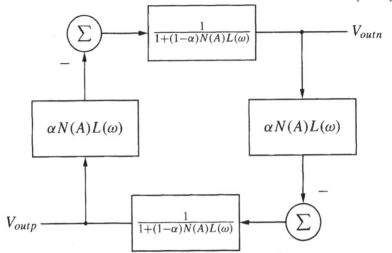

Figure A.1: Simplified block schematic of figure 4.8

Eliminating $\Delta\omega$ yields :

$$\left[\left(\left.\frac{\partial M}{\partial\omega}\right|_{A_0,\omega_0}\right)^2 + \left(M\left.\frac{\partial\phi}{\partial\omega}\right|_{A_0,\omega_0}\right)^2\right]\Delta\sigma$$

$$= \left[M\left.\frac{\partial\phi}{\partial\omega}\right|_{A_0,\omega_0}\left.\frac{\partial M}{\partial A}\right|_{A_0,\omega_0} - M\left.\frac{\partial\phi}{\partial A}\right|_{A_0,\omega_0}\left.\frac{\partial M}{\partial\omega}\right|_{A_0,\omega_0}\right]\Delta A \qquad \text{(A.8)}$$

For a limit cycle to be stable, a positive increment ΔA requires a positive $\Delta\sigma$ to compensate the amplitude growth and a negative increment ΔA requires a negative $\Delta\sigma$. A necessary condition for stability of the limit cycle thus would be that the sign of $\Delta A/\Delta\sigma$ is positive, or since the amplitude M is always positive :

$$\left.\frac{\partial\phi}{\partial\omega}\right|_{A_0,\omega_0}\left.\frac{\partial M}{\partial A}\right|_{A_0,\omega_0} - \left.\frac{\partial\phi}{\partial A}\right|_{A_0,\omega_0}\left.\frac{\partial M}{\partial\omega}\right|_{A_0,\omega_0} > 0 \qquad \text{(A.9)}$$

2. Polar Form of the Coupled Open Loop Transfer Function

The simplified block schematic of the coupled SOPA is depicted in figure A.1. From this schematic the open loop transfer function can be easily derived :

$$TF = \frac{(\alpha N(A)L(\omega))^2}{(1 + (1-\alpha)N(A)L(\omega))^2} = TF_1^2 \qquad \text{(A.10)}$$

Since the non-linear, amplitude dependent parts are very hard to separate from the frequency dependent parts, a graphical analysis is hard to perform. To facilitate the stability analysis, (A.10) is written in polar coordinates. Therefor the loop filters transfer function is also written in polar coordinates :

$$L(\omega) = \left(M^*(\omega)\exp(j\phi^*)\right)^n = M^*(\omega)^n \exp(jn\phi^*) \tag{A.11}$$

$$M^*(\omega) = \frac{\omega_c}{\sqrt{\omega_c^2 + \omega^2}} \tag{A.12}$$

$$\phi^*(\omega) = \arctan\left(-\frac{\omega}{\omega_c}\right) \tag{A.13}$$

This gives for (A.10) :

$$TF_1 = M(A,\omega)\exp(j\phi(A,\omega)) \tag{A.14}$$

$$M(A,\omega) = \frac{\alpha M^*(\omega)^n N(A)}{\sqrt{(1-\alpha)^2 N(A)^2 M^*(\omega)^{2n} + 2(1-\alpha)N(A)M^*(\omega)^n \cos(n\phi^*(\omega)) + 1}} \tag{A.15}$$

$$\phi(A,\omega) = n\phi^*(\omega) - \arctan\left(\frac{(1-\alpha)N(A)M^*(\omega)^n \sin(n\phi^*(\omega))}{1 + (1-\alpha)N(A)M^*(\omega)^n \cos n\phi^*(\omega)}\right) \tag{A.16}$$

3. Calculation of the Stability Conditions

The stability of the in-phase oscillation (A_0, ω_0) and the counter-phase oscillation (A_1, ω_1) needs to be determined by filling in the values of the respective limit cycle amplitudes and frequencies in the stability criterion (A.9).

$$\omega_0 = \omega_1 = \omega_c \tan\left(\frac{\pi}{n}\right) \tag{A.17}$$

$$A_0 = \frac{2V_{DD}}{\pi}\cos^n\left(\frac{\pi}{n}\right) \tag{A.18}$$

$$A_1 = \frac{2V_{DD}}{\pi}(1-2\alpha)\cos^n\left(\frac{\pi}{n}\right) \tag{A.19}$$

The following observations can be made in advance for the values of the loop filter parameters in the limit cycle operating points :

$$\phi^*(\omega)\Big|_{\substack{\omega_0 \\ \omega_1}} = -\frac{\pi}{n} \tag{A.20}$$

$$M^*(\omega)\Big|_{\substack{\omega_0 \\ \omega_1}} = \cos\left(\frac{\pi}{n}\right) \tag{A.21}$$

$$\tag{A.22}$$

When filling in the non-linearity $N(A) = 2V_{DD}/A$, the following evaluations hold :

$$N(A)\big|_{A_0} = \cos^{-n}\left(\frac{\pi}{n}\right) \quad \Rightarrow \quad N(A)M^*(\omega)^n\big|_{A_0} = 1 \tag{A.23}$$

$$N(A)\big|_{A_1} = \frac{\cos^{-n}\left(\frac{\pi}{n}\right)}{1-2\alpha} \quad \Rightarrow \quad N(A)M^*(\omega)^n\big|_{A_0} = \frac{1}{1-2\alpha} \tag{A.24}$$

Furthermore the partial derivatives of the main components of (A.15) and (A.16) can be easily calculated and evaluated. Note that for compactness of notation the explicit dependency of

$N(A)$, $M^*(\omega)$ and $\phi^*(\omega)$ is omitted:

$$\left.\frac{\partial N}{\partial}\right|_{A_0} = -\frac{\pi}{2V_{DD}}\cos^{-2n}\left(\frac{\pi}{n}\right) \tag{A.25}$$

$$\left.\frac{\partial N}{\partial}\right|_{A_1} = -\frac{\pi}{2V_{DD}(1-2\alpha)^2}\cos^{-2n}\left(\frac{\pi}{n}\right) \tag{A.26}$$

$$\left.\frac{\partial \phi^*}{\partial \omega}\right|_{\substack{\omega_0 \\ \omega_1}} = \frac{-\cos^2\left(\frac{\pi}{n}\right)}{\omega_c} \tag{A.27}$$

$$\left.\frac{\partial M^*}{\partial \omega}\right|_{\substack{\omega_0 \\ \omega_1}} = \frac{-\cos^3\left(\frac{\pi}{n}\right)}{\omega_c} \tag{A.28}$$

For the calculation of the stability criterion, the chain rule is heavily used. In this way, early simplifications can be introduced in the calculation.

$$\frac{\partial \phi}{\partial A} = \frac{\partial \phi}{\partial N}\frac{\partial N}{\partial A} \tag{A.29}$$

$$= \frac{-\left((1-\alpha)M^*\sin(n\phi^*)\right)}{(1+(1-\alpha)NM^*\cos(n\phi^*))^2 + ((1-\alpha)NM^*\sin(n\phi^*))^2}\frac{\partial N}{\partial A} \tag{A.30}$$

Since (A.20):

$$\sin(n\phi^*)\big|_{\substack{\omega_0 \\ \omega_1}} = 0 \tag{A.31}$$

and

$$\left.\frac{\partial N}{\partial A}\right|_{\substack{A_0 \\ A_1}} \neq 0 \tag{A.32}$$

the stability criterion (A.9) is reduced to

$$\left.\frac{\partial M}{\partial A}\frac{\partial \phi}{\partial \omega}\right|_{A_x,\omega_x} > 0 \tag{A.33}$$

For calculating the first part of (A.33), the chain rule is used :

$$\frac{\partial M}{\partial A} = \frac{\partial M}{\partial N}\frac{\partial N}{\partial A} \tag{A.34}$$

with

$$\frac{\partial M}{\partial N} = \frac{\alpha M^{*n}\sqrt{(1-\alpha)^2N^2M^{*2n} + 2(1-\alpha)NM^{*n}\cos(n\phi^*) + 1}}{(1-\alpha)^2N^2M^{*2n} + 2(1-\alpha)NM^{*n}\cos(n\phi^*) + 1}$$
$$\frac{\alpha NM^{*n}\dfrac{N(1-\alpha)^2M^{*2n}+(1-\alpha)M^{*n}\cos(n\phi^*)}{\sqrt{(1-\alpha)^2n^2M^{*2n}+2(1-\alpha)nM^{*n}\cos(n\phi^*)+1}}}{-\frac{}{(1-\alpha)^2N^2M^{*2n} + 2(1-\alpha)NM^{*n}\cos(n\phi^*) + 1}} \tag{A.35}$$

Since ϕ^* evaluates to $-\pi/n$ for both limit cycle solutions, (A.35) can be evaluated to

$$\left.\frac{\partial M}{\partial N}\right|_{\phi^*=\frac{-\pi}{n}} = \frac{\alpha M^{*n}\left(\left|(1-\alpha)NM^{*n} - 1\right| - N(1-\alpha)M^{*n}\left|(1-\alpha)NM^{*n} + 1\right|\right)}{\left((1-\alpha)NM^{*n} - 1\right)^2} \tag{A.36}$$

Evaluating this expression in the limit cycle solutions gives :

$$\left.\frac{\partial M}{\partial N}\right|_{A_0,\omega_0} = \frac{\cos^n\left(\frac{\pi}{n}\right)}{\alpha} \tag{A.37}$$

$$\left.\frac{\partial M}{\partial N}\right|_{A_1,\omega_1} = \frac{(1-2\alpha)^2\cos^n\left(\frac{\pi}{n}\right)}{\alpha} \tag{A.38}$$

The only term that has to be calculated is the derivative of the phase to the frequency :

$$\frac{\partial\phi}{\partial\omega} = \frac{\partial\phi}{\partial\phi^*}\frac{\partial\phi^*}{\partial\omega} + \frac{\partial\phi}{\partial M^*}\frac{\partial M^*}{\partial\omega} \tag{A.39}$$

$$\frac{\partial\phi}{\partial\phi^*} = n - n\frac{(1-\alpha)NM^{*n}}{\left(1+(1-\alpha)NM^{*n}\cos(n\phi^*)\right)^2 + \left((1-\alpha)NM^{*n}\sin(n\phi^*)\right)^2} \times$$
$$\left[\cos(n\phi^*)\left(1+(1-\alpha)NM^{*n}\cos(n\phi^*)\right)\right.$$
$$\left. + \left((1-\alpha)NM^{*n}\sin(n\phi^*)\right)\sin(n\phi^*)\right] \tag{A.40}$$

This can be further simplified by filling in $\phi^* = -\pi/n$:

$$\left.\frac{\partial\phi}{\partial\phi^*}\right|_{\phi^*=\frac{-\pi}{n}} = n\left[1 + \frac{(1-\alpha)NM^{*n}}{1-(1-\alpha)NM^{*n}}\right] \tag{A.41}$$

Evaluating (A.41) for the two possible solutions gives :

$$\left.\frac{\partial\phi}{\partial\phi^*}\right|_{A_0,\omega_0} = \frac{n}{\alpha} \tag{A.42}$$

$$\left.\frac{\partial\phi}{\partial\phi^*}\right|_{A_1,\omega_1} = \frac{n}{\alpha}(2\alpha-1) \tag{A.43}$$

The derivative of the phase to the loop filters magnitude can be calculated as :

$$\frac{\partial\phi}{\partial M^*} = -\frac{(1-\alpha)NnM^{*(n-1)}\sin(n\phi)}{\left(1+(1-\alpha)NM^{*n}\cos(n\phi)\right)^2 + \left((1-\alpha)NM^{*n}\sin(n\phi)\right)^2} \tag{A.44}$$

Due to the absence of $\sin(n\phi^*)$ in the nominator, the evaluation to $\phi^* = -\pi/n$ becomes :

$$\left.\frac{\partial\phi}{\partial M}\right|_{\phi^*=\frac{-\pi}{n}} = 0 \tag{A.45}$$

Filling (A.34) and (A.39) in (A.33) and taking (A.45) into account, gives the following stability condition :

$$\left.\left(\frac{\partial M}{\partial N}\frac{\partial N}{\partial A}\frac{\partial\phi}{\partial\phi^*}\frac{\partial\phi^*}{\partial\omega}\right)\right|_{A_x,\omega_x} > 0 \tag{A.46}$$

Evaluating (A.46) for the in-phase solution, means substituting the different partial derivatives with (A.27), (A.27), (A.37) and (A.25)

$$\frac{n\pi\cos^{(2-n)}\left(\frac{\pi}{n}\right)}{2\alpha^2 V_{DD}\omega_c} > 0 \tag{A.47}$$

For the counter-phase case (A.28), (A.28), (A.38) and (A.26) has to be filled in, in (A.46), giving :

$$\frac{n(1-2\alpha)\pi\cos^{(2-n)}\left(\frac{\pi}{n}\right)}{2\alpha^2 V_{DD}\omega_c} > 0 \tag{A.48}$$

References

[Abramowitz and Stegun, 1972] Abramowitz, Milton and Stegun, Irene A. (1972). *Handbook of mathematical functions: with formulas, graphs, and mathematical tables*. Dover New York (N.Y.), 9th print edition.

[ADTT1-6,] ADTT1-6. *ADTT1-6 RF Transformer datasheet*. mini-circuits, http://www.minicircuits.com/cgi-bin/spec?cat=tranfrmr&model= ADTT1-6&pix%=cd636.gif&bv=4.

[Allstot, 1982] Allstot, David J. (1982). A precision variable-supply CMOS comparator. *IEEE J. Solid-State Circuits*, 17(6):1080–1087.

[Annema et al.,] Annema, Anne-Johan, Geelen, Govert, and de Jong, Peter. 5.5 V tolerant I/O in a 2.5V 0.25 μm cmos.

[Aparicio and Hajimiri, 2002] Aparicio, Roberto and Hajimiri, Ali (2002). Capacity limits and matching properties of integrated capacitors. *IEEE J. Solid-State Circuits*, 37(3):384–393.

[Atherton, 1975] Atherton, D.P. (1975). *Nonlinear Control Engineering*. Van Nostrand Reinhold.

[Babanezhad, 1998] Babanezhad, J.N. (1998). A 100MHz 50 ω -45dB 3.3V CMOS line-driver for ethernet and fast ethernet networking applications. In *ISSCC Digest of Technical Papers*. IEEE.

[Baker et al., 1998] Baker, Jacib R., Li, Harry W., and Boyce, David E. (1998). *CMOS, circuit design, layout and simulation*, chapter 26 : nonlinear analog circuits, pages 685–717. series on microelectronics. IEEE press.

[Barkhausen, 1935] Barkhausen, H (1935). *Lehrbuch der Elektronen-Rohre*, chapter 3. Ruckkopplung. Verlag S. Hirzel, Leipzig.

[Baze, 1991] Baze, Mel (1991). two novel fully complementary self-biased CMOS differential amplifiers. *IEEE J. Solid-State Circuits*, 26(2):165–168.

[Benton et al., 2001] Benton, Roger, Apfel, Russell, Webb, Bruce, Wenske, Jerome, Schopfer, Walt, and Thiel, Frank (2001). A high-voltage line driver (HVLDR) for combined voice and data services. In *ISSCC Digest of Technical Papers*, pages 302–303.

[Bergen et al., 1982] Bergen, Arthur R., Chua, Leon O., Mees, Alistair I., and Szeto, ellen W. (1982). Error bounds for general describing function problems. *IEEE Trans. Circuits Syst.*, CAS-29(6):345–354.

[Bicakci et al., 2003] Bicakci, Ara, Kim, Chun-Sup, Lee, Sang-Soo, and Conroy, Cormac (2003). A 700 mW CMOS line driver for ADSL central office applications. In *ISSCC Digest of Technical Papers*, volume 46, pages 414–415. IEEE.

[Blackmore, 1981] Blackmore, Denis (1981). The describing function for bounded nonlinearities. *IEEE Trans. Circuits Syst.*, cas-28(5):442–447.

[Candy and Temes, 1992] Candy, James C. and Temes, Gabor C (1992). *Oversampling Delta-sigma data convertors, Theory, design and simulation*, chapter Oversampling methods for A/D and D/A conversion. IEEE Press.

[Casier et al., 1998a] Casier, H., Wouters, P., Graindourze, B., and Sallaerts, D. (1998a). a 3.3 V, low-distortion ISDN line driver with a novel quiescent current control circuit. *IEEE J. Solid-State Circuits*, 32(7):1130–1133.

[Casier et al., 1998b] Casier, H., Wouters, P., and Sallaerts, D. (1998b). A 3.3-v low-distortion ISDN line driver with a novel quiescent current control circuit. *IEEE J. Solid-State Circuits*, 33(7):1130–1133.

[Casson, 1910] Casson, Herbert N. (1910). *The History of The Telephone*. A.C. McClurg & Co, Chicago, first edition.

[Cloetens, 2001] Cloetens, Leon (2001). Broadband access : The last mile. In *ISSCC Digest of Technical Papers*, pages 18–21.

[Conroy et al., 1999] Conroy, Cormac, Sheng, Samuel, Feldman, Arnold, Uehara, Gregory, Yeung, Alfred, Hung, Chih-Jen, Subramanian, Vivek, Chiang, Patrick, Lai, Paul, Si, Xiaomin, Fan, Jerry, Flynn, Denis, and He, Meiqing (1999). A CMOS analog front-end IC for DMT ADSL. In *ISSCC Digest of Technical Papers*, pages 240–241.

[Cornil et al., 1999a] Cornil, J.P., Sevenhans, J., Spruyt, P., Mielants, M., and Braet, S. (1999a). DMT ADSL and VDSL circuits and systems. In *AACD 99 workshop on advances in analog circuit design*, pages 1–17.

[Cornil et al., 1999b] Cornil, J.P., Z.Y., Chang, Louagie, F., W., Overmeire, and J., Verfaille (1999b). A 0.5μm CMOS ADSL analog front-end IC. In *ISSCC Digest of Technical Papers*, pages 238–239.

[Cresi et al., 2001] Cresi, M., Hester, R., Maclean, K., Agah, M., Quarfoot, J., Kozak, C., Gibson, N., and Hagen, T. (2001). An ADSL central office analog front-end integrating actively-terminated line driver, receiver and filters. In *ISSCC Digest of Technical Papers*, pages 304–305.

[Cypkin et al., 1962] Cypkin, Ja. Z., Guileminet, J., and Gille, J.-C. (1962). *Théorie des asservissements par plus-ou-moins*. Dunod Paris.

[Dallago, 1997] Dallago, Enrico (1997). Advances in high-frequency power conversion by delta-sigma modulation. *IEEE Trans. Circuits Syst. I*, 44:712–721.

[De Graeve, 2002] De Graeve, Frank (17/04/2002). Internet is geen hype meer. *De Standaard*.

[Dixon, 1999] Dixon, Lloyd H (1999). *Magnetics Design for Switching Power Supplies.* http://ti-training.com/courses/coursedetail.asp?iCSID=1152.

[Eaton, 2002] Eaton, John W. (2002). *GNU Octave Manual.* Network Theory Limited.

[Frenzel, 2001] Frenzel, Louis E. (2001). DSL and cable battle for broadband supremacy in The Last Mile. *Electronic Design,* 49(15):58–64.

[Gelb and Vander Velde, 1968] Gelb, Arthur and Vander Velde, Wallace (1968). *Multiple-input Describing Functions and Nonlinear System Design.* McGraw-Hill Book Company.

[Gibson and Sridhar, 1963] Gibson, J.E. and Sridhar, R. (1963). A new dual-input describing function and an application to the stabiolity of forced oscillations. *Trans. AIEE, pt II, Apll. Ind.,* pages 65–70.

[Green and Williams, 1992] Green, Timothy C. and Williams, Barry W. (1992). Spectra of delta-sigma modulated inverters: An analytical treatment. *IEEE Trans. Power Electronics,* 7(4):644–654.

[Hindmarsh, 1983] Hindmarsh, A. C. (1983). *Scientific Computing,* chapter ODEPACK, A systematic Collection of ODE Solvers, pages 55–64. North-Holland, Amsterdam.

[In-Stat/MDR, 2002] In-Stat/MDR (2002). Reports about the death of broadband are premature - subscriber growth remains robust. http://www.instat.com/newmk.asp?ID=278.

[Ingels et al., 2002] Ingels, Mark, Bojja, Sena, and Wouters, Patrick (2002). A 0.5 μm CMOS low-distortion low-power line driver with embeded digital adaptive bias algorithm for integrated ADSL analog front-ends. In *ISSCC Digest of Technical Papers,* pages 324–325.

[Iwai, 1999] Iwai, Hiroshi (1999). CMOS technology - year 2010 and beyond. *IEEE J. Solid-State Circuits,* 34(3):357–366.

[Jacobsen, 1999] Jacobsen, Krista (1999). VDSL: The next step in the DSL progression. In *DSPS Fest 99,* http://www.ti.com/sc/docs/general/dsp/fest99/telecom/2jacobsen.pdf. Texas Instruments.

[Kalet, 1989] Kalet, Irving (1989). The multitone channel. *IEEE Trans. Commun.,* 37(2):119–124.

[Kappes, 2000] Kappes, Michael S. (2000). A 3 V CMOS low-distortion class AB line driver suitable for HDSL applications. *IEEE J. Solid-State Circuits,* 35(3):371–376.

[Khorramabadi, 1992] Khorramabadi, Haideh (1992). A CMOS line driver with 80-dB linearity for ISDN applications. *IEEE J. Solid-State Circuits,* 27(4):539–544.

[Laaser et al., 2001] Laaser, P., Eichler, T., Wenske, H., Herbison, D., and Eichfeld, H. (2001). A 285 mW CMOS single chip analog front end for G.SHDSL. In *ISSCC Digest of Technical Papers,* pages 298–299.

[Lienhard, 2000] Lienhard, John H. (2000). *The Engines of Our Ingenuity – An Engineers Looks at Technology and Culture.* Oxford University Press.

[Lindgren, 1964] Lindgren, Allen G. (1964). Limit cycles in symmetric multivariable systems. *IEEE Trans. Automat. Contr.,* 9:119–120.

[Luke, 1962] Luke, Yudell L. (1962). *Integrals of Bessel functions*. McGraw-Hill.

[MacLachlan, 1955] MacLachlan, N.W. (1955). *Bessel Functions for Engineers*. Oxford university press London, second edition.

[Maclean et al., 2003] Maclean, Kenneth, Corsi, Marco, Hester, Richard, Quarfoot, James, Melsa, Peter, Halbach, Robert, Kozak, Carmen, and Hagan, Tobin (2003). A 620mW zero-overhead class G full-rate ADSL central-office line driver. In *ISSCC Digest of Technical Papers*, volume 46, pages 412–413. IEEE.

[Mahadevan and Johns, 2000] Mahadevan, Rajeevan and Johns, David (2000). A differential 160MHz self-terminating adaptive CMOS line driver. In *ISSCC Digest of Technical Papers*, volume 43, pages 436–437.

[Maxwell, 1996] Maxwell, Kim (1996). Asymmetric digitil subscriber line : Interim technology for the next forty years. *IEEE Commun. Mag.*, pages 100–106.

[May et al., 2001] May, Marcus W., May, Michael R., and Willis, John E. (2001). A synchronous dual-input switching dc-dc converter using multibit noise-shaped switch control. In *ISSCC Digest of Technical Papers*, pages 358–359.

[Mertens and Steyaert, 2002] Mertens, Koen L.R. and Steyaert, Michiel S.J. (2002). A 700 MHz 1 W fully differential CMOS Class-E power amplifier. *IEEE J. Solid-State Circuits*, 37(2):137–141.

[MestDagh et al., 1993] MestDagh, D.J.G., Spruyt, P.M.P., and Biran, B. (1993). Effect of amplitude clipping in DMT-ADSL transceivers. *Electronics Letters*, 29(15):1354–1355.

[Midcom 50702R,] Midcom 50702R. *Midcom 50702R digital transformer datasheet*. Midcom, http://www.midcom-inc.com/digital/pdf/AlcatelMicroelectronicsADSL.pdf.

[Moons, 2003] Moons, Elvé (2003). looking to/for low power ADSL drivers in the DSLAM. In *Proceedings Workshop on Advances in Analog Circuit Design*.

[Moyal et al., 2003] Moyal, M., Groepl, M., Werker, H., Mitteregger, G., and Schambacher, J. (2003). A 700/900 mW/channel CMOS dual analog front-end IC for VDSL with integrated 11.5/14.5dBm line drivers. In *ISSCC Digest of Technical Papers*, volume 46, pages 416–417. IEEE.

[Nauta and Dijkstra, 1998] Nauta, Bram and Dijkstra, Marcel B. (1998). Analog video line driver with adaptive impedance matching. In *ISSCC Digest of Technical Papers*, volume 41, pages 318–319.

[Philips et al., 1999] Philips, K., van den Homberg, J., and Dijkmans, E.C. (1999). PowerDAC : a single chip audio DAC with a 70% efficient powerstage in 0.5μm - cmos. In *ISSCC Digest of Technical Papers*. IEEE.

[Pierdomenico et al., 2002] Pierdomenico, John, Wurcer, Scott, and Day, Bob (2002). A 744 mW adaptive supply full-rate ADSL CO driver. In *ISSCC Digest of Technical Papers*, pages 320–321.

[Piessens and Steyaert, 2001] Piessens, Tim and Steyaert, Michiel (2001). SOPA : A high-efficiency line driver in 0.35μm CMOS using a self-oscillating power amplifier. In *ISSCC Digest of Technical Papers*, pages 306–307.

[Piessens and Steyaert, 2002a] Piessens, Tim and Steyaert, Michiel (2002a). A central officed combined ADSL-VDSL line driver in .35 μm CMOS. In *Proceedings Custom Integrated Circuits Conference*, pages 45–48. IEEE.

[Piessens and Steyaert, 2002b] Piessens, Tim and Steyaert, Michiel (2002b). Design considerations and experimental verification of a self oscillating line driver in .35 μm CMOS. In *Proceedings European Solid-State Circuits Conference*, pages 783–786. IEEE.

[Piessens and Steyaert, 2003a] Piessens, Tim and Steyaert, Michiel (2003a). Highly efficient xDSL line drivers in 0.35 μm CMOS using a self-oscillating power amplifier. *IEEE J. Solid-State Circuits*, 38(1):22–29.

[Piessens and Steyaert, 2003b] Piessens, Tim and Steyaert, Michiel (2003b). Oscillator pulling and synchronisation issues in self-oscillating class D power amplifiers. In *Proceedings European Solid-State Circuits Conference*. IEEE.

[Powel, 1970] Powel, M. J. D. (1970). *Numerical Methods for Nonlinear Algebraic Equations*, chapter A Hybrid Method for Nonlinear Equations. Gordon and Breach.

[Rabaey, 1996] Rabaey, Jan M. (1996). *Digital Integrated Circuits, A Design Perspective*. Prentice-Hall, Inc.

[Rabaey, 2002] Rabaey, Jan M. (2002). Signal integrity in SOC - challenges and solutions. In *Electrical Issues in SoC/SoP Design, Satellite Workshop, ESSSDERC-ESSCIRC 2002*, pages 31–51.

[Rohde and Schwartz,] Rohde and Schwartz. *I/Q Modulation Generator R&S AMIQ - Data sheet*. Rohde and Schwartz, http://www.rohde-schwarz.com/www/datsheet. nsf/file/AMIQ_23_web.pdf, pd 0757.3970.23 edition.

[Roza, 1997] Roza, Engel (1997). Analog-to-digital conversion via duty-cycle modulation. *IEEE Trans. Circuits Syst. II*, 44(11):907–914.

[Sabouti and Shariatdoust, 2002] Sabouti, Faramarz and Shariatdoust, Reza (2002). A 740 mW ADSL line driver for central office with 75 dB MTPR. In *ISSCC Digest of Technical Papers*, pages 322–323.

[Sæther et al., 1996] Sæther, Trond, Hung, Chung-Chih, Qi, Zheng, and Aaserud, Oddvar (1996). High speed, high linearity CMOS buffer amplifier. *IEEE J. Solid-State Circuits*, 31(2):255–258.

[Sands et al., 1999] Sands, Nicholas P., Naviasky, Eric, Evans, William, Mengele, Martin, Faison, Kevin, Frost, Craig, Casas, Michael, and Williams, Michelle (1999). An integrated analog front-end for VDSL. In *ISSCC Digest of Technical Papers*, pages 246–247.

[Sevenhans et al., 2002] Sevenhans, Jan, De Wilde, Wim, Moons, Elve, Reusens, Peter, Berti, Laurent, Kiss, Lajos, Casier, Herman, and Gielen, George (2002). Driving the DSL highway : high speed, high density, low power, low cost. In *Proceedings European Solid-State Circuits Conference*, pages 555–562.

[Slotine and Li, 1991] Slotine, Jean-Jacques E. and Li, Weiping (1991). *Applied Nonlinear Control*. Prentice-Hall, Inc.

[Sowlati and Leenaerts, 2002] Sowlati, Tirdad and Leenaerts, Domine (2002). A 2.4 GHz 0.18μm CMOS self-biased cascode power amplifier with 23 dBm output power. In *ISSCC Digest of Technical Papers*, pages 294–295. IEEE Press.

[Starr et al., 1999] Starr, Thomas, Cioffi, John, M., and Silverman, Peter J. (1999). *Understanding Digital Subscriber Line Technology*. Prentice-Hall, Inc.

[Su and McFarland, 1998] Su, David K. and McFarland, William J. (1998). An IC for lineaizing RF power amplifiers using envelope elimination and restoration. *IEEE J. Solid-State Circuits*, 33(12):2252–2258.

[Tellado-Mourelo, 1999] Tellado-Mourelo, Jose (1999). *Peak to Average Power Reduction for Multicarrier Modulation*. PhD thesis, Departement of Electrical Engineering, Stanford University.

[van der Zee and van Tuijl, 1998] van der Zee, R. A. R. and van Tuijl, A. J. M. (1998). A power efficient audio amplifier combining switching and linear techniques. In *Proceedings of the 24th European Solid-State Circuits Conference*, pages 288–291.

[Wambacq and Sansen, 1998] Wambacq, Piet and Sansen, Willy (1998). *Distortion analysis of analog integrated circuits*. Kluwer Academic Publishers.

[Wang, 2001] Wang, Qi (2001). Draft trial-use standard t1e1.4, very-high-bit-rate digital subscriber line (VDSL) metallic interface. part 1: Functional requirements and common specification. ftp://ftp.t1.org/pub/t1e1/E1.4/DIR2000/0e140093.PDF.

[Weisstein,] Weisstein, Eric W. Eric weisstein's world of mathematics (mathworld^{TM}). http://mathworld.wolfram.com.

[Zojer et al., 2000] Zojer, Bernhard, Koban, Rüdger, Pichler, Joachim, and Paoli, Gerhard (2000). A broadband high-voltage SLIC for a splitter and transformerless combined ADSL-Lite/POTS linecard. *IEEE J. Solid-State Circuits*, 35(12):1976–1987.

[Zojer et al., 1997] Zojer, Bernhard, Koban, Rüdiger, Petchacher, Reinhard, and Sereinig, Wolfgang (1997). A 150-v subscriber line interface circuit (SLIC) in a new BiCMOS/DMOS-technology. *IEEE J. Solid-State Circuits*, 32(9):1475–1480.

Index